MATHEMATICS RESEARCH DEVELOPMENTS

AN INTRODUCTION TO FUZZY SETS

Mathematics Research Developments

Additional books and e-books in this series can be found on Nova's website under the Series tab.

MATHEMATICS RESEARCH DEVELOPMENTS

AN INTRODUCTION TO FUZZY SETS

VALDEMAR F. ANDERSEN
EDITOR

nova
science publishers
New York

Library of Congress Cataloging-in-Publication Data

Names: Andersen, Valdemar F., editor. Title: An introduction to fuzzy sets / Valdemar F. Andersen, editor.
Description: New York : Nova Science Publishers, [2020] | Series: Mathematics research developments | Includes bibliographical references and index. | Summary: "An Introduction to Fuzzy Sets provides a comparison of the quality of life in urban, intermediate and rural NUTS III regions in Portugal, with the main goal of identifying and analysing the necessary and conditions for a high quality of life in those different regions. The authors assess the necessary and sufficient conditions for higher Human Development Index levels, aiming to determine whether the same pattern could be used to explain the happiness index. In order to represent the applications of fuzzy set theory as well as neuro-fuzzy in industry, a literature review of these topics is carried out. As some researchers have efficiently utilized fuzzy logic and neuro-fuzzy, in-depth discussions are provided for stimulating future investigations. Following this, using the L. Zadeh theory of fuzzy sets, the authors consider all types of uncertainties in oil fields and oil production to make a decision as to what model is best in such a fuzzy environment. Additionally, several challenges are explored, such as: the fuzzy random finite difference numerical method, possibilistic uncertainty modeling, and the development of a fuzzy Wilks' theorem to model the hybrid structure of randomness and fuzziness modeling. In closing, a standard fuzzy arithmetic method is used for solving fuzzy equations, as well as for the optimization of fuzzy objectives. The fuzzy variable of the equation is fuzzified using a fuzzy set"-- Provided by publisher.
Identifiers: LCCN 2020020544 (print) | LCCN 2020020545 (ebook) | ISBN 9781536180121 (paperback) | ISBN 9781536180138 (adobe pdf)
Subjects: LCSH: Quality of life--Portugal. | Quality of life. | Fuzzy sets. | Social sciences--Mathematics.
Classification: LCC HN594 .I47 2020 (print) | LCC HN594 (ebook) | DDC 306.09469--dc23
LC record available at https://lccn.loc.gov/2020020544
LC ebook record available at https://lccn.loc.gov/2020020545

Published by Nova Science Publishers, Inc. † New York

Contents

PREFACE

An Introduction to Fuzzy Sets provides a comparison of the quality of life in urban, intermediate and rural NUTS III regions in Portugal, with the main goal of identifying and analysing the necessary and conditions for a high quality of life in those different regions.

The authors assess the necessary and sufficient conditions for higher Human Development Index levels, aiming to determine whether the same pattern could be used to explain the happiness index.

In order to represent the applications of fuzzy set theory as well as neuro-fuzzy in industry, a literature review of these topics is carried out. As some researchers have efficiently utilized fuzzy logic and neuro-fuzzy, in-depth discussions are provided for stimulating future investigations.

Following this, using the L. Zadeh theory of fuzzy sets, the authors consider all types of uncertainties in oil fields and oil production to make a decision as to what model is best in such a fuzzy environment.

Additionally, several challenges are explored, such as: the fuzzy random finite difference numerical method, possibilistic uncertainty modeling, and the development of a fuzzy Wilks' theorem to model the hybrid structure of randomness and fuzziness modeling.

In closing, a standard fuzzy arithmetic method is used for solving fuzzy equations, as well as for the optimization of fuzzy objectives. The fuzzy variable of the equation is fuzzified using a fuzzy set.

Chapter 1 - This study provides a comparison of the Quality of Life (QoL) in urban, intermediate and rural NUTS III regions in Portugal. The main goal is to identify and analyse the necessary and sufficient conditions for high levels of QoL in those different regions. Those areas are distinguished based on the population in rural areas (population density), share of population in rural areas and size of urban centres. The authors use data provided by several institutions, through Statistics Portugal and PORDATA. The Gini Index is the proxy used for QoL, i.e., the higher the Gini Index, the lower the global development and consequently QoL. The domains of QoL studied are: (i) living material conditions; (ii) productive activity; (iii) health; (iv) education; (v) leisure and social interactions; (vi) physical security; (vii) governance; and (viii) natural living conditions. In order to obtain these domains, the authors use exploratory factor analysis. Having the domains as conditions, fuzzy sets are applied in order to evaluate the necessary and sufficient conditions in urban and rural regions. The authors' results point to some similar and other different necessary and sufficient conditions for rural and urban regions; for example, a higher Gini Index is related to a less ageing population and access to health services, both in rural and urban regions.

Chapter 2 - The Human Development Index (HDI) has been an indicator used to evaluate various aspects of human development since the 1990s. Subject to several criticisms, namely the fact that it does not measure some kinds of relevant variables, the HDI has undergone several transformations in its calculation. In parallel, other indices have been used to measure other situations. For example, the World Happiness Report (WHR) is an annual publication of the United Nations Sustainable Development Solutions Network, created to measure a happiness index, whose assumptions are different. The objective of this research is to identify the necessary and sufficient conditions for higher HDI levels and to determine whether the same pattern can be used to explain the happiness index. The questions under study are: "Is the HDI mostly driven by economic forces? In the end, do happiness and human development have the same basis? Which are the Life and Economic necessary and sufficient conditions to reach high levels of HDI and WHI?"

Chapter 3 - Artificial intelligence, also named machine intelligence is the emulation of human intelligence procedures using machines, particularly computer systems. It is a branch of computer science, which purposes to generate intelligent machines and has become a fundamental part of the technology industry. Artificial intelligence makes it feasible for machines to acquire knowledge from experience, adapt to new inputs and accomplish human like tasks. In the last decades, artificial intelligence has had significant economic and societal benefits. Artificial intelligence methodologies such as fuzzy logic and neuro-fuzzy are broadly employed in various applications for example in business, marketing, control engendering, health care, and social services. In order to represent the applications of fuzzy set theory as well as neuro-fuzzy in industry, besides to provide a basis for future research, a literature review of these topics in industry is carried out in this book chapter. As some researchers have efficiently utilized fuzzy logic and neuro-fuzzy, indepth discussions are provided for stimulating future investigations.

Chapter 4 - The oil field and oil production belong to the class of the big, complex and hierarchical systems. Therefore, there is a need for good modeling. Today there are many means of modeling. But they all can consider only one concrete type of uncertainty. Using the L. Zadeh theory of fuzzy sets the authors can consider all types of uncertainties at the same time and to make a decision of what model is the best in such fuzzy environment. As result the authors can get up the more intellectually control and monitoring of system of oil and gas production.

Chapter 5 - TOKAMAK is a type of fusion reactor that use magnetic force to control plasma, fourth state of matter in which a separation between electrons and neutrons takes place by a pulsed high heat flux of magnitude 2×10^5 MW/m^2. In TOKAMAK, heat flux gets injected into the plasma facing components due to plasma disruption. Melting and evaporation both takes place in the first wall of TOKAMAK type of fusion reactor due to which high heat load plasma facing components of the wall cracks. Ignorance of uncertainty (in this situation uncertainty is known as a mixture of probabilistic and epistemic uncertainty) is associated with the materials used for the design. Several researchers have performed experiments to study the melting and evaporation behaviour of the first wall of fusion reactor. In order to understand the initiation of melting and evaporation of plasma facing components, the first wall of a fusion blanket is modeled as 2-D rectangular slab with the surface facing plasma disruption by an applied heat flux. The rear surface of the slab is assumed as cooled by convection. The relevant parameters affecting the heat transfer during the early phases of heating as well as for large times are taken into account as uncertain due to their fuzziness and randomness. Fuzziness and randomness consistency principle is applied to model the uncertainty of the thermal response of the first wall. Fuzzy random set concept is very new to this kind of uncertainty modeling and therefore, this challenges the design of plasma facing components subject to simultaneous melting and evaporation. Finite Difference (Forward Time Central Space) based numerical solution with fuzzy random variables of the relevant parameters of the governing heat transfer equations is applied to obtain the temporal and spatial variation of temperature. The interpretation of results obtained are two folds: (a) Randomness uncertainty of the temperature of melting and evaporation of first wall of the plasma facing components is occurred at any alpha cut of the fuzziness, whereas (b) fuzziness uncertainty of the same components due to heat load is addressed at 95% confidence level by 2.5^{th} and 97.5^{th} percentile of the temperature. Water and Helium both are considered for cooling of the system. This chapter presents several challenges such as: (a) fuzzy random finite difference numerical method; (b) possibilistic uncertainty modeling; and (c) development of Fuzzy Wilks' theorem to model hybrid structure of randomness and fuzziness modeling.

Chapter 6 - A standard fuzzy arithmetic method is used for solving fuzzy equations and optimization of fuzzy objectives. The fuzzy variable of the equation is fuzzified using a fuzzy set. A fuzzy set is defined as a membership function with a degree of membership varying from zero to one where zero represents non-membership, one represents full membership and values between zero and one represent partial membership. The vertical axis of the fuzzy set is decomposed into several intervals. Each interval is solved separately using the interval theory. The resultant intervals are recomposed according to degree of membership and resultant fuzzy set is obtained. The resultant fuzzy sets are defuzified to get the crisp values. The fuzzy arithmetic method is applied to a case study of the reservoir operation in which fuzzy storage continuity and fuzzy soil moisture balance equations are solved and fuzzy model objective of the minimizing crops moisture deficits is obtained. The method is found quite effective and easy to apply in solving complex equations.

In: An Introduction to Fuzzy Sets
Editor: Valdemar F. Andersen
ISBN: 978-1-53618-012-1

Chapter 1

QUALITY OF LIFE: URBAN VERSUS RURAL ANALYSIS BASED ON FUZZY SETS APPROACH

Andreia Dionísio[1] and Conceição Rego[2,*]
[1]Cefage, Department of Management, Universidade de Évora, Évora, Portugal
[2] Cefage, Department of Economics, Universidade de Évora, Évora, Portugal

ABSTRACT

This study provides a comparison of the Quality of Life (QoL) in urban, intermediate and rural NUTS III regions in Portugal. The main goal is to identify and analyse the necessary and sufficient conditions for high levels of QoL in those different regions. Those areas are distinguished based on the population in rural areas (population density), share of population in rural areas and size of urban centres. We use data provided by several institutions, through Statistics Portugal and PORDATA. The Gini Index is the proxy used for QoL, i.e., the higher the Gini Index, the

* Corresponding Author's E-mail: andreia@uevora.pt.

lower the global development and consequently QoL. The domains of QoL studied are: (i) living material conditions; (ii) productive activity; (iii) health; (iv) education; (v) leisure and social interactions; (vi) physical security; (vii) governance; and (viii) natural living conditions. In order to obtain these domains, we use exploratory factor analysis. Having the domains as conditions, fuzzy sets are applied in order to evaluate the necessary and sufficient conditions in urban and rural regions. Our results point to some similar and other different necessary and sufficient conditions for rural and urban regions; for example, a higher Gini Index is related to a less ageing population and access to health services, both in rural and urban regions.

Keywords: quality of life, rural regions, urban regions, fuzzy-sets, exploratory factor analysis

Introduction

Improving the living conditions and well-being of populations in different countries is a priority for societies. Identification of the factors that allow a better quality of life, in the diverse cultural, economic, social, political and geographic contexts, is crucial so that public and private decision-makers can act more effectively. For a long time, assessment of nations' progress was based almost exclusively on analysis of the evolution of GDP. The intensity of GDP growth was considered the most appropriate way to learn about economic evolution and, therefore, reflect well-being. However, it was revealed that economic growth does not, inevitably, translate into better levels of development and QoL. Observation of the economic and social situation in different geographies showed that it was possible to have more growth and more development but also more growth without better levels of development and quality of life. Thus, it becomes essential to disconnect the concepts of growth and development. While the first measures countries and regions' capacity to create more wealth, the second implies that the wealth created is distributed more evenly across society. Growth becomes only a necessary condition for development, but not a sufficient condition for improved QoL.

This distinction began in the 1990s at the United Nations with the creation of the Human Development Index. The Human Development Index (HDI) is a comparative measure used to classify countries by their degree of "human development," distinguishing them as developed (very high human development), developing (medium and high human development) and underdeveloped (low human development). The index, developed by the economists Amartya Sen and Mahbub ul Haq, is calculated by combining three dimensions: A long and healthy life (Life expectancy at birth), access to knowledge (Average Years of Study and Expected Years of Schooling) and a decent standard of living (GDP (PPS) per capita).

In addition to the United Nations, other international organizations such as the OECD and the European Union currently devote much attention to effective measurement of quality of life. In 2018, the OECD published a report entitled "Beyond GDP: Measuring What Counts for Economic and Social Performance," by the OECD High Level Expert Group on the Measurement of Economic Performance and Social Progress, coordinated by Joseph E. Stiglitz, Jean -Paul Fitoussi and Martine Durand, where attention is drawn to the quality and relevance of data and information from which public decision-makers make their respective decisions: "Metrics matter for policy and policy matters for well-being" (OECD, 2018). Finding good indicators that can reveal and measure the true situation of countries and their population, in addition to GDP, is fundamental in building knowledge: "If you don't measure important phenomena, you may not act on them" (OECD, 2018: 33). In order to produce relevant information to assess citizens' effective quality of life, the OECD has been developing the Better Life Initiative. "The OECD Better Life Initiative focuses on developing statistics to capture aspects of life that matter to people and that shape the quality of their lives; this allows for a better understanding of what drives the well-being of people and nations, and what needs to be done to achieve greater progress for all." The Better Life Index is based on 11 topics: community, education, environment, civic engagement, health, housing, income, jobs, life satisfaction, safety and work-life balance.

In line with this are the studies promoted by the European Commission (EC), through EUROSTAT, with the aim of providing a set of indicators that monitor social and environmental progress. "The point of this broader perspective is to measure whether, and, if so, how our societies are progressing towards inclusiveness and ensuring wellbeing for our generation without compromising the ability of future generations to meet their own needs." From the perspective of EUROSTAT, QoL is assessed through a multiplicity of variables, with objective and subjective characteristics, in 8 + 1 dimensions: Material living conditions, Productive or other main activity, Health, Education, Leisure and social interactions, Economic security and physical safety, Governance and basic rights, Natural and living environment and Overall experience of life.

With the same concern, the Well-being Index is available in Portugal, with data for the country (NUTS I). This indicator is produced by INE (Statistics Portugal) and aims to monitor the evolution of well-being and social progress in two instrumental domains: households' material living conditions and quality of life. This concept of well-being is very close to that proposed by EUROSTAT, since it "covers not only material living conditions, but also other factors which contribute to quality of life, namely those related to the environment, good health, proper educational attainment, use of time, in particular work-life balance, the value of living in a society, a good level of democratic participation as well as access to and involvement in cultural and leisure activities." This index covers the following domains: economic well-being; economic vulnerability; labour and earnings; health; work-life balance; education; knowledge and skills; social interactions and subjective well-being; civic participation and governance; personal security; environment. According to current QoL perspectives, countries and regions are more developed not only when they create more wealth, but especially when they are more just, inclusive, environmentally, socially and economically sustainable and when they allow citizens to have more balanced lives.

As is well known, there are significant differences in development between the countries and regions of the European Union. These differences are marked between different countries as well as within countries themselves (Portugal is an obvious case of these differences). Since the 1980s, these differences have justified major European investment in the context of the structural and cohesion policy. The territorial organization of countries and regions focuses on cities and urban areas and the different relations they establish with rural areas. Urban and rural areas differ through the prevailing economic functions, social groups, lifestyles and landscapes (Rego et al., 2017). Traditionally, urban areas are described as dynamic, young, innovative, with higher levels of quality of life, while rural areas are described as depopulated, aging, with low levels of economic activity, employment and quality of life. The researcher's initial idea is that, today as before, urban areas are perceived as more developed, providing better quality of life for citizens, as opposed to rural areas. However, studies analysing the differences in QoL between urban and rural areas do not, in general, show such conclusive results. Sucksmith et. al. (2009), based on data from the European Quality of Life Survey (EQLS 2003), find that "the richest countries in the EU show little evidence of significant urban-rural differences, whereas, in the poorer countries of the east and south, rural areas have a much lower level of perceived welfare and quality of life (…). Despite this, subjective well-being is not significantly different" Sucksmith et. al. (2009: 1275). Carta et al. (2012) examine the association between subjective QoL in rural or urban areas, in six Italian regions. The authors include the variables of age and gender in the analysis. They find that "men show a higher subjective quality of life than women; subjective quality of life decreases with age in both genders and men are more sensitive to urban/rural residence than women" (Carta et al., 2012:169).

In a scenario where the majority of Europeans live in urban areas ["about half (51%) of Europeans live in a medium to large town (25%) or in a city or city suburb (26%), while 49% live in the open countryside, a village or small town" and where "since 2007, there has been a decrease in the number of people living in the open countryside and an increase in those living in medium to large towns" (Eurofound, 2014: 6)], a policy brief finds that "there are complex patterns in terms of differences within rural and within urban areas, with inequalities pointing in different directions for the various quality of life indicators" (Eurofound, 2014: 7). For example, "about one in three (34%) people in the EU28 who live in a city or city suburb live in a one-person household compared with one in four (23%) in the open countryside; in rural areas, 50% of single households are retired people compared with 37% in urban areas; in rural areas, more single households (37%) are women aged 60+ than in urban areas (27%)"; (...) "People in urban areas have higher incomes, but this does not mean less deprivation and less difficulty making ends meet, probably due to the higher cost of living and greater inequality in cities" (Eurofound, 2014: 6). This pattern is also present in analysis of the Portuguese case. Amado et al. (2019), based on a study comparing the Quality of Life (QoL) of cities that lost population with cities that gained population, concluded that, despite the heterogeneity in cities that shrunk, this set presents, on average, higher QoL than others that have grown. Apparently, inhabitants' attraction to cities does not seem to be associated with QoL; the relationship with factors that promote, strictly, greater economic competitiveness could be more relevant (cities that grow are mainly in urban and metropolitan areas). In a study about the differences in QoL between urban and rural regions, using fsQCA methodology, Rego et- al (2017) did not find an unequivocal relationship between better QoL and the predominance of an urban or rural environment. This study concludes that QoL (where the proxy is Purchasing Power) results from the combination of factors associated with higher density in urban areas (low levels of illiteracy, absence of burned areas, total companies, waste selectively collected) with others that preserve rural characteristics (Longevity Index and Living in a rural area). These results probably reflect the effects of the European cohesion policy which, over the last 30 years, has improved the QoL across the country, mostly through supplying local public services and the infrastructure network.

From another point of view and with an approach that links civic participation and citizens' political options to rural/urban areas and global development levels are the recent studies by Rodriguez-Posé (2018) and Dijkstra et. al. (2019). Rodriguez-Posé (2018) discussed the "revenge of places that don't matter," relating populist political choices to people living in less developed territories. In fact, there seems to be a connection between political options in urban areas (more democratic and progressive), with greater economic dynamism, and less dynamic rural areas (more conservative and populist). Similarly, the study by Dijkstra et. al. (2019), analysing the fact that support for parties opposed to European Union (EU) integration has risen rapidly, finds that the "anti-EU vote is mainly a consequence of local economic and industrial decline in combination with lower employment and a less educated workforce" (Dijkstra et. al., 2019:1).

With this research we intend to discuss the necessary and sufficient conditions for high levels of QoL in rural, intermediate and urban Portuguese NUTS III regions. The analysis is based on a set of variables, available in official Portuguese statistics, in the domains of (i) living material conditions; (ii) productive activity; (iii) health; (iv) education; (v) leisure and social interactions; (vi) physical security; (vii) governance; and (viii) natural living conditions. The proxy for QoL is the Gini Index. The fuzzy sets qualitative comparative analysis is used in order to reach the objetives.

After this introduction, the paper continues by presenting the data and describing the methodology in section 2. In section 3 the results are discussed and the fourth section concludes.

METHODS

In this section we present the data collected and the methods used to analyse them and reach the main objective of this research: to identify the causal conditions for quality of life (QoL), investigating and comparing the necessary and sufficient conditions for that outcome between rural, intermediate and urban regions.

The classification of NUTS III regions as "predominantly rural," "intermediate" or "predominantly urban" is according to the EUROSTAT proposal. The classification is obtained through 3 steps: i) identifying the population in rural areas: "rural areas" are all areas outside an urban cluster (an urban cluster has a density of at least 300 inhabitants per km^2 and a minimum population of 5 000 residents); ii) classification of regions based on the share of population in rural areas: "Predominantly rural" if the share of the population living in rural areas is higher than 50%; "Intermediate" if the share of the population living in rural areas is between 20% and 50% and

"Predominantly urban" if the share of the population living in rural areas is below 20%; iii) the size of urban centres in the region: a predominantly rural region containing an urban centre of more than 200 000 inhabitants making up at least 25% of the regional population becomes intermediate; an intermediate region containing an urban centre of more than 500 000 inhabitants making up at least 25% of the regional population becomes predominantly urban.

The selection of domains of analysis and variables to be used to assess QoL was inspired by the work carried out by the European Commission (EC) "GDP and beyond." However, this EC study is based on the organization and collection of data for the different EU countries, meaning it is not possible to replicate the same variables at the NUTS III level (used in this study). The variables were selected from those available in the statistical data for NUTS III in Portugal, through Statistics Portugal and PORDATA (cf. Table 1). In this EC study, the assessment of global satisfaction, in terms of quality of life, designated as Overall experience of life, is obtained through the variables of Life satisfaction and Affects, which are available only for countries and within the scope of the study. As the literature in this field draws attention to the gaps associated with using GDP as a measure of QoL, in this article the outcome condition will be the Gini Index. The Gini Index is an indicator of inequality in income distribution, which aims to summarize the asymmetry of that distribution in a single value, assuming values between 0 (when all individuals have equal income) and 100 (when all income is concentrated in a single individual. This variable is proposed as a proxy for the global QoL measure, since it is considered that levels of inequality in income distribution are higher in cases where countries and regions' global development is lower. The inclusion of this variable is fundamental, since the methodology implies the existence of a dependent variable.

Table 1. Domains and variables of Quality of Life

Domains	Variables	Code
Living material conditions	Purchasing power per capita (2017); Index (number) - %	PP
	Average household size, according to the Census (2011); Individual - Mean	HouSize
Productive or other main activity	Employees by level of education: Upper-secondary and post-secondary non-tertiary (2017); Individual	EmplSec
	Employees by level of education: Higher education (2017); Individual	EmplHigher
	Employment sex ratio, according to the Census, and by sector of economic activity: primary (2011); Ratio - %	EmplSectorI
	Employment sex ratio, according to the Census, by sector of economic activity: secondary (2011); Ratio - %	EmplSectorII
	Employment sex ratio, according to the Census, by sector of economic activity: tertiary (2011); Ratio - %	EmplSectorIII
	Inactive population per 100 active people, according to the Census, by sex (Males) (2011); Ratio - %	InactiveMale
	Inactive population per 100 active people, according to the Census, by sex (Females) (2011); Ratio - %	InactiveFemale
	Unemployment, according to the Census, by highest level of education completed: upper secondary (2011); Individual	UnempUpperSec
	Unemployment, according to the Census, by highest level of education completed: middle level (2011); Individual	UnempMiddle
	Unemployment, according to the Census, by highest level of education completed: higher education (2011); Individual	UnempHEduc
	Employment, according to the Census: primary sector	EmplCSI
	Employment, according to the Census: secondary sector	EmplCSII
	Employment, according to the Census: tertiary sector	EmplCSIII

Table 1. (Continued)

Domains	Variables	Code
Health	Inhabitants per doctor (2018); Ratio	Doctor
	Ageing Index (2018); Ratio %.	Ageing
Education	Resident population aged 15 and over, according to the Census, by highest level of educational qualifications obtained: Upper-secondary (2011); Individual	Educ1_UpperSec
	Resident population aged 15 and over, according to the Census, by highest level of educational qualifications obtained: middle level (2011); Individual	Educ2_Middle
	Resident population aged 15 and over, according to the Census, by highest level of educational qualifications obtained: higher education (2011); Individual	Educ3_HEduc
	Teaching staff in pre-school, primary, lower secondary and upper-secondary education: total (2018); Individual	Teaching
	Illiteracy rate, according to the Census: total	Illiteracy
	Pupils enrolled in pre-school, primary, lower secondary and upper-secondary education: total and by level of education	Students
Leisure	Cinema: audience per thousand inhabitants (2018); Rate - per thousand	Cinema
	Town Council expenditure on culture and sports as a % of total expenditure (2018); Proportion - %	Town
	Live shows: audience per thousand inhabitants (2018); Rate - per thousand	Live_Shows
Security	Crimes registered by police per thousand inhabitants (2018); Ratio	Security
Governance	Abstention rate in parliamentary elections(2019); Rate - %	Gov
Natural Living Conditions	Small towns (2018); Number of small towns	SmallTown
	Urban waste selective collection per inhabitant (2018); kg/ inhab. - Ratio	UrbanWaste
	Resident population in places with 10 thousand inhabitants and over, according to the Census (2011); Individual	Inahbi

Source: Statistics Portugal and PORDATA.

The Gini Index data for Portuguese NUTS III (Figure 1) show that levels of inequality are relatively close in the different regions under analysis, varying between 30 and 35%. The maximum values of inequality are registered in the metropolitan areas of Lisbon and Porto – where there are also higher levels of per capita income. The autonomous island regions of the Azores and Madeira also have high values. It should be noted that the Azores are among the poorest Portuguese regions and Madeira is one of the regions with the highest GDPpc (this region is a duty free area). These results, concerning interregional balance and relatively low asymmetry, are closely related to the Portuguese income redistribution policy, which supports the disposable income of the poorest individuals, and to the effects of the cohesion policy, which promotes the supply of public services in the domains of education, health, infrastructure, culture and others.

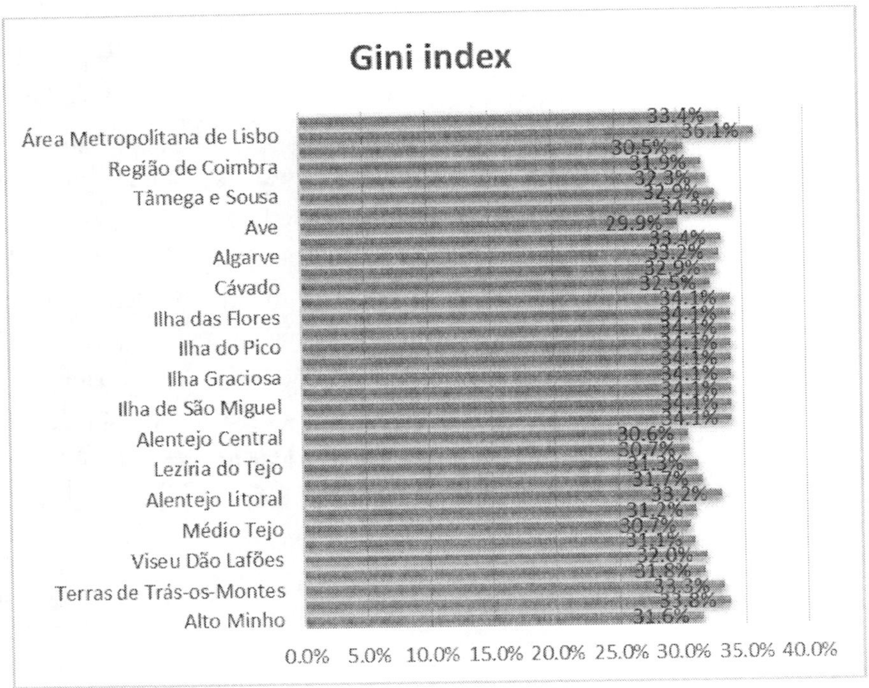

Figure 1. Gini index for the set of regions under study.

The data were collected from Statistics Portugal (INE) and PORDATA and refer to the last year available (in some cases, 2017 or 2018); for many variables, the data are only available based on the last Portuguese population CENSUS (2011).

Given the number of variables and the significant linear correlation between variables in the same domain, a factor analysis procedure was applied in order to reduce the number of independent variables to include in the fuzzy-sets analysis. The global objective of factor analytical techniques is to find a way to summarise the information contained in a set of original variables into a smaller set of new composite dimensions (factors) with a minimum loss of information (Hair et al., 2014).

In terms of methods, this study is mostly based on fuzzy-set qualitative comparative analysis (fsQCA), a methodology which evaluates how conditions cause a specific outcome (Wagemann & Schneider 2010). Selection of this methodology is related to the objectives of the research and the potentialities of fsQCA. While quantitative methodologies aim to find cause-effect relations between dependent and independent variables (regression analysis for example), fsQCA looks for the combinations of conditions forming the basis in terms of necessary and sufficient conditions for a particular outcome Vis (2012). "A causal condition is called necessary if the instances of the outcome constitute a subset of the instances of the causal condition" (Ragin, 2006, p. 297). The quality of the necessary condition is evaluated through two main concepts: consistency and coverage. Schneidee, Schulze-Bentrop and Paunescu (2010) consider that coverage can be read as a measure of triviality, that is, the greater the coverage, the less the triviality. The authors consider non-trivial conditions whose coverage is greater than 60%. Equation 1 shows how to calculate the coverage rate.

$$Coverage\ (Yi \leq Xi) = \frac{\sum_{i=1}^{I} min(xi,yi)}{\sum_{i=1}^{I} xi} \qquad (1)$$

In the probabilistic versions of fsQCA, consistency is used to quantify the percentage of observations that confirm the rule. The measure of consistency of Ragin (2006, 2008) measures the extent of cases that do not observe the defined rule and gives greater weight to higher failures than to small failures (this is because the system is not binary, but calibrated).

In terms of mathematical definition, the consistency rate of causal condition X is defined by:

$$Consistency\ (Yi \leq Xi) = \frac{\sum_{i=1}^{I} min(xi,yi)}{\sum_{i=1}^{I} yi} \tag{2}$$

A sufficient condition exists if the result or outcome (Y) occurs whenever condition X is present. In this case we say that X is a sufficient condition for Y (Ragin, 2006). To obtain sufficient conditions, it is necessary to create and model the "truth table," which is based on consistency and the combination of conditions, because an isolated condition X is rarely a sufficient condition for Y. The cut-off consistency for the truth table is set at 0.8 (Ragin, 2006). The quality of the obtained results is based on two main measures:

i) solution coverage - percentage of Y cases explained by the model, including all combinations, Equation 3.

$$Solutioncoverage(Xi \leq Yi) = \frac{\sum_{i=1}^{I} min(xi,yi)}{\sum_{i=1}^{I} yi} \tag{3}$$

ii) Solution Consistency: percentage in which the solution belongs to a subset of the result, Equation 4.

$$SolutionConsistency(Xi \leq Yi) = \frac{\sum_{i=1}^{I} min(xi,yi)}{\sum_{i=1}^{I} xi} \tag{4}$$

Since the contribution of Ragin (1987), it has undergone several developments (Ragin, 2008).

To the best of our knowledge, no study compares the necessary and sufficient conditions of QoL between rural and urban regions. Probably because the concept of QoL is quite broad and subjective, and also because the available data is generally insufficient for traditional quantitative studies. This is also the reason for using fsQCA in this research. In this field, we may refer to the study by Paykani, Rafiey and Saijadi (2018), which analysed the structural conditions that explain life expectancy at birth using fsQCA. The results point to a configuration of conditions including a high level of governance, education, wealth and a well-developed health system as being consistently sufficient for high life expectancy. Tho (2015) studied the human capital and quality of work life of marketers, based on fsQCA and structural equation modelling, finding some divergences in both results and concluding that hope and optimism are sufficient conditions for the outcome condition analysed.

A major potential of fsQCA lies in this methodology being able to analyze different combinations of conditions in a given problem (Ragin, 2008). Another interesting feature is that it is a suitable methodology even when samples are small or medium-sized (Vis, 2012). This is the case here, since we study 33 regions of Portugal, in 3 groups: (i) Urban group with 7 observations; (ii) Intermediate group with 4 regions, and; (iii) Rural group with 22 observations.

Applying fsQCA implies calibrating variables, since it does not use original data (as, for example, regression-based methodologies do). The calibration process is fundamental to the methodology and implies, according to Ragin (2000), 'substantive and theoretical knowledge relevant to set membership.' Normally, in the calibration process, the researcher identifies for all conditions and for the outcome, the fully in set (the condition/outcome will have the value of 1), the fully out set (0 value), and also the crossover point (0.5), meaning that the observation is neither in nor out regarding the set. With calibration, the conditions and the outcome are rescaled and range from 0 to 1. We base our calibration on the percentile approach. According to Ragin (2008), this approach is appropriate when we have continuous data. Given this, the 'fully in' set is defined as the 95th percentile, the 'fully out' as the 5th percentile and the 'neither in nor out' point was defined by the median (50th percentile). The same criterion is used for all conditions and the outcome under study. The results were estimated using the current version of the fs/QCA software package (2.5).

RESULTS

This section presents the results obtained in this study. In the first part, the factor analysis results, emphasizing the interpretation of each factor obtained. In the second part, the fsQCA results for each group of regions.

Factor Analysis

As mentioned in Section 2, the domains of QoL are formed of several variables, measured in different units, which would make the causal analysis and respective interpretation difficult. The dimension of our database was reduced and analysed using factor analysis (procedure Factor in SPSS version 24). The KMO index and the Bartlett's test for sphericity were used to confirm the robustness and quality of the factor obtained.

The analysis was performed using principal components with Varimax rotation and component extraction was based on the eigenvalue criterion. The results for each domain (or dimension) are presented in Tables 2, 3 and 4.

The KMO index and the Bertlett's test for the domain of *Living material conditions*, *Health* and *Leisure* did not show evidence of robustness and quality of the information extracted. Given the extremely low quality of the factors extracted, we decided to use the original variables.

Concerning *Productive or other main activity*, the results are presented in Table 2.

We emphasise in Table 2 variables with a coefficient greater than 0.6. making it possible to distinguish three components related to productive or other main activity: the first component associated with production and employment related to the level of education (Prod_Educ), the second component associated with the inactive population (Pro_Inactive) and the third component related to production and employment related to sector (Prod_Sectors). A similar procedure was followed for the *Education* domain, where the results point to the existence of one component, which we called Education (see Table 3).

Table 2. Relationship between variables and different components after rotation for Productive or other main activity

Domain: Productive or other main activity	**Component**		
EmplSec	0.987	-0.075	-0.064
EmplHigher	0.976	-0.062	-0.023
EmplSectorI	-0.203	-0.113	0.756
EmplSectorII	-0.324	-0.472	0.333
EmplSectorIII	0.115	-0.04	0.768
InactiveMale	-0.114	0.939	-0.179
InactiveFemale	-0.294	0.893	0.052
UnempUpperSec	0.991	-0.064	-0.061
UnempMiddle	0.988	-0.07	-0.085
UnempHEduc	0.99	-0.053	-0.058
EmplCSI	0.555	0.254	-0.424
EmplCSII	0.906	-0.093	-0.146
EmplCSIII	0.986	-0.057	-0.066
KMO = 0.765			
Bartlett's Test = 985.235 **			

Note: ** 1% significance level.

Table 3. Relationship between variables and different components after rotation for *Education*

Domain: Education	**Education**
Educ1_UpperSec	0.995
Educ2_Middle	0.995
Educ3_HEduc	0.993
Illiteracy	-0.388
Students	0.99
KMO = 0.81	
Bartlett's Test = 485,632 **	

Note: ** 1% significance level.

Table 4. Relationship between variables and different components after rotation for *Natural living conditions*

Domain: Natural living conditions	Natural
SmallTown	0.928
Inahbi	0.928
KMO = 0.65	
Bartlett's Test = 223.084 **	

Note: ** 1% significance level.

Table 4 presents the results of factor analysis for the *natural living conditions* domain. The results point to one component, which we call Natural, as results of the linear combination between the variables of SmallTown and Inahbi.

Causal Analysis

To perform the causal analysis, we start by studying the necessary conditions for QoL for the three groups: rural, intermediate and urban. It is important to remember that in this paper, the proxy for QoL is the Gini index, which means that higher values of this index indicate a higher concentration of wealth, which means a lower quality of life for the whole population.

The necessary conditions for QoL (and the absence of Gini) for the three groups are presented in Table 5.

The relevant necessary conditions, those with consistency levels above 0.8, are not the same for the three types of regions. For example, if we focus on ~Gini (absence of Gini index) we find that for rural regions the necessary conditions are the absence of the household size (~HouSize), the absence of employment in sectors I and III (~ProdSectors), the presence of ageing, the absence of abstention (~Gov) and the absence of natural living conditions (~Natural). For intermediate regions, the absence of inhabitants per doctor (Doctor) is the only relevant necessary condition for ~Gini.

Table 5. Necessary conditions for QoL for rural, intermediateand urban regions

Conditions	Rural				Intermediate				Urban			
	Outcome: Gini		Outcome: ~Gini		Outcome: Gini		Outcome: ~Gini		Outcome: Gini		Outcome: ~Gini	
	Cons	Cov	Cons	Cov	Cons	Cov	Cons	Cov	Cons	Cov	Cons	Cov
PP	0.499	0.564	0.606	0.601	0.745	0.731	0.365	0.330	0.743	0.714	0.607	0.604
~PP	0.647	0.652	0.560	0.495	0.317	0.351	0.703	0.718	0.588	0.591	0.713	0.742
HouSize	0.765	0.815	0.388	0.363	0.433	0.489	0.672	0.701	0.616	0.605	0.609	0.620
~HouSize	0.402	0.428	0.803	0.750	0.736	0.708	0.510	0.454	0.613	0.603	0.612	0.623
ProdEducation	0.433	0.462	0.767	0.716	0.462	0.480	0.646	0.620	0.758	0.835	0.445	0.507
~ProdEducation	0.734	0.782	0.424	0.396	0.635	0.660	0.458	0.440	0.552	0.490	0.855	0.786
ProdInnactive	0.442	0.465	0.768	0.709	0.413	0.448	0.703	0.703	0.649	0.628	0.516	0.517
~ProdInnactive	0.723	0.781	0.419	0.397	0.726	0.726	0.448	0.413	0.501	0.500	0.629	0.650
ProdSectors	0.702	0.819	0.368	0.376	0.649	0.754	0.448	0.480	0.740	0.695	0.521	0.506
~ProdSectors	0.465	0.456	0.823	0.708	0.553	0.520	0.771	0.670	0.473	0.488	0.686	0.732
Doctor	0.615	0.765	0.414	0.452	0.721	0.833	0.443	0.472	0.433	0.466	0.722	0.805
~Doctor	0.559	0.521	0.784	0.641	0.543	0.514	0.844	0.736	0.819	0.740	0.521	0.487
Ageing	0.380	0.419	0.799	0.772	0.428	0.486	0.641	0.672	0.445	0.461	0.663	0.711
~Ageing	0.794	0.818	0.399	0.361	0.712	0.682	0.510	0.452	0.720	0.674	0.496	0.481
Education	0.486	0.530	0.628	0.601	0.471	0.485	0.620	0.589	0.715	0.794	0.462	0.531
~Education	0.634	0.660	0.509	0.465	0.601	0.631	0.458	0.444	0.578	0.509	0.821	0.749
Cinema	0.397	0.473	0.697	0.729	0.452	0.472	0.635	0.613	0.776	0.780	0.496	0.517
~Cinema	0.772	0.744	0.495	0.419	0.630	0.652	0.453	0.433	0.519	0.499	0.789	0.785
Town	0.573	0.567	0.752	0.653	0.380	0.457	0.656	0.728	0.494	0.494	0.749	0.776
~Town	0.649	0.749	0.501	0.507	0.774	0.709	0.510	0.432	0.776	0.749	0.511	0.511

Conditions	**Rural**				**Intermediate**				**Urban**			
	Outcome: Gini		Outcome: ~Gini		Outcome: Gini		Outcome: ~Gini		Outcome: Gini		Outcome: ~Gini	
	Cons	Cov	Cons	Cov	Cons	Cov	Cons	Cov	Cons	Cov	Cons	Cov
LiveShows	0.462	0.527	0.680	0.680	0.524	0.542	0.573	0.547	0.677	0.679	0.477	0.495
~LiveShows	0.719	0.719	0.527	0.462	0.563	0.588	0.521	0.503	0.496	0.478	0.690	0.689
Security	0.698	0.796	0.428	0.428	0.558	0.655	0.521	0.565	0.791	0.806	0.469	0.495
~Security	0.498	0.498	0.796	0.698	0.630	0.587	0.682	0.587	0.504	0.478	0.816	0.802
Gov	0.796	0.878	0.330	0.319	0.659	0.662	0.510	0.473	0.601	0.578	0.663	0.662
~Gov	0.382	0.394	0.874	0.790	0.476	0.513	0.635	0.632	0.649	0.651	0.577	0.599
Natural	0.317	0.368	0.831	0.848	0.553	0.599	0.599	0.599	0.776	0.782	0.533	0.556
~Natural	0.869	0.854	0.381	0.329	0.630	0.630	0.599	0.553	0.560	0.537	0.791	0.785

Note: Cons – Consistency rate, Cov – Coverage rate.

For urban regions, the results point to the absence of Education, absence of crimes registered by police per thousand inhabitants (~Security) and absence of natural living conditions (~Natural). If the absence of crimes is considered consensual, the absence of education for ~Gini is not a common result for higher levels of quality of life. In urban regions, the results point to safety and natural living conditions. This may be due to a growing concern about the safety of goods and people, which is more common in urban environments. Natural living conditions is common to both types of regions, revealing this to be a current concern for the population. In fact, the results obtained in verifying the necessary conditions, for the 3 types of regions, are generally in accordance with the factors identified as contributing to the existence of low difference levels of income asymmetry (Cf. Figure 1).

We continue our analysis by considering the sufficient conditions. We follow the procedure proposed by Ragin and Fiss (2008), which shows the intermediate solutions but identifying core conditions (represented by larger symbols) and peripheral conditions (represented by smaller symbols). Tables 6, 7 and 8 present the sufficient conditions for QoL for rural, intermediate and urban regions.

The results in Table 6 point to the existence of 5 combinations of sufficient conditions. The unique coverage (UC) values reveal the importance of each combination obtained, and the overall solution consistency (OSC) presents a value of 0.9216, indicating the level of belonging of the combination as a subset of the result. In these combinations of sufficient conditions for the Gini index in rural regions, we highlight the absence of Ageing and Natural Living Conditions, the presence of abstention (Gov) and, in the first two combinations, the presence of the inhabitants per doctor variable. We may assume that the combinations revealed are sufficient for increasing the concentration of wealth. The most traditional determinants of income inequality are related to level of education, participation in the labour market, age, region of residence and sector of activity, among others (Rodrigues et. al., 2012). Therefore, higher inequality in income distribution in rural areas highlights traditionally urban characteristics. However, we must take into consideration that rural regions have a set of characteristics that distinguish them from others: in general, rural regions are characterized by an aging population, valorisation of environmental conditions related to lower occupation density, i.e., depopulation and low levels of entrepreneurial initiatives, by the difficulty in accessing health services and other collective services and infrastructure (Amado et al. (2019), Rego et. al. (2017), Ferrão (2002), Eurofound (2014), Unidade de Missão para a Valorização do Interior (2016)) and by different participation in elections (more conservative) (Rodriguez-Posé (2018), Dijkstra et. al. (2019).

Table 6. Sufficient conditions for higher levels of Gini Index for rural regions

	1	**2**	**3**	**4**	**5**
ProdEducation	⊗	⊗	⊗	⊗	•
ProdInnactive	⊗	⊗	⊗	⊗	•
Doctor	●	●	⊗	⊗	⊗
Ageing	⊗	⊗	⊗	⊗	⊗
Education	⊗	⊗	•	⊗	⊗
Cinema	⊗	⊗	⊗	•	⊗
Security	●	⊗	•	•	•
Natural	⊗	⊗	⊗	⊗	•
Gov		●	●	●	●
RC	0.3754	0.2662	0.2354	0.2167	0.1476
UC	0.1433	0.0571	0.0281	0.0503	0.0307
Cons.	0.9600	0.9454	0.9616	0.9584	1.000
OSCov.	0.6416				
OSC	0.9216				

Legend: Cons: consistency; RC: Raw coverage; UC: Unique coverage; OSC: Overall Solution consistency; OSCov: Overall Solution coverage. ●Core causal condition (presence). •Peripheral causal condition (presence). ⊗Core causal condition (absent). ⊗Peripheral causal condition (absent).

Table 7 presents the combinations of sufficient conditions for intermediate regions. As can be seen, these combinations are globally antagonistic, except in the purchasing power (PP) condition. It is worth noting the low value of global coverage, which indicates that the percentage of cases explained by the model, including all combinations, is around

55.76%. It should be noted that this analysis has only 4 observations, a fact that may lessen the strength of the analysis. fsQCA has the great advantage of being a suitable method for small samples, but even so, the small size and variability of this sample may be factors that detract from the robustness of the analysis carried out.

Table 7. Sufficient conditions for higher levels of Gini Index for intermediate regions

	1	2
PP	●	●
ProdEducation	⊗	•
ProdInnactive	⊗	•
Doctor	•	⊗
HouSize	•	⊗
Education	⊗	•
Cinema	⊗	•
Security	⊗	•
Natural	⊗	•
Gov	⊗	•
RC	0.3269	0.2884
UC	0.2692	0.2307
Cons.	0.9855	1.000
OSCov.	0.557692	
OSC	0.991453	

Legend: Cons: consistency; RC: Raw coverage; UC: Unique coverage; OSC: Overall Solution consistency; OSCov: Overall Solution coverage. ●Core causal condition (presence). •Peripheral causal condition (presence). ⊗Core causal condition (absent). ⊗Peripheral causal condition (absent).

Table 8 shows the results obtained for urban regions. The two combinations obtained have a global coverage of 45.80%, a value considered low, although the overall consistency is 98.9%. In these combinations, the absence of Ageing and Inhabitants per doctor (Doctor) should be highlighted. These results are in line with characteristics prevailing in urban areas. The combinations obtained are difficult to interpret, given the high number of conditions forming them, registering antagonistic positions in some of the conditions, namely PP, ProdEducation, ProdInnactive, Education, Security, Natural and Gov. In fact, urban regions are generally characterized by the existence of higher incomes, more resident and active population and higher levels of education. However, we admit that these results may be associated with the fact that the differences between Portuguese regions are not very significant, in most of the dimensions evaluated, since most of them are related to the cohesion policy (ADC, 2018). The major differences between Portuguese regions are in the domains of competitiveness while the cohesion policy promotes the least divergence between regions (ADC, 2018).

Table 8. Sufficient conditions for higher levels of Gini Index for urban regions

	1	2
PP	⊗	•
ProdEducation	⊗	•
ProdInnactive	⊗	•
Doctor	⊗	⊗
HouSize	•	⊗
Education	⊗	•
Ageing	⊗	⊗
Town	⊗	⊗
Security	⊗	•
Natural	⊗	•
Gov	•	⊗
RC	0.1933	0.3231
UC	0.1348	0.2646
Cons.	0.9743	1.000
OSCov.	0.4580	
OSC	0.9890	

Legend: Cons: consistency; RC: Raw coverage; UC: Unique coverage; OSC: Overall Solution consistency; OSCov: Overall Solution coverage. ●Core causal condition (presence). •Peripheral causal condition (presence). ⊗Core causal condition (absent). ⊗Peripheral causal condition (absent).

CONCLUSION

It is interesting to analyse the conditions that promote QoL, since nowadays wealth is increasingly concentrated and analysing well-being through the indicator of GDP per capita has proved to be insufficient to capture the diversity of domains and economic, social and environmental influences on individuals' well-being in the different places where they live. International institutions such as the United Nations, the OECD, the European Commission or the World Bank pay close attention to this issue and work on indicators that make it possible to assess and compare quality of life globally in different countries and regions.

Studies carried out in the field of comparing QoL in urban and rural regions conclude that the factors promoting well-being in these two types of regions are different, and it is not possible to affirm that levels of QoL are higher in regions where economic growth is higher.

In this study the proxy for QoL is the Gini Index, i.e., the higher the Gini Index, the lower the level of global development and QoL We apply fsQCA and the main findings are the follows: i) the sufficient conditions are robust for the Gini index in rural regions: the absence of Ageing and Natural Living Conditions, the presence of abstention (Gov) and the presence of the inhabitants per doctor variable; ii) in intermediate regions standing out among sufficient conditions are the effect of purchasing power on the existence of high levels of Gini Index; iii) in the case of urban regions, the sufficient conditions highlight the absence of Ageing and Inhabitants per doctor (Doctor). The nature of the outcome condition certainly influences the results obtained, since it is a variable of income distribution. In summary, a higher Gini Index is related to a less ageing population and access to health services, in both rural and urban regions.

This methodology is particularly suitable to analyse this kind of issue since the data sample is small and the need to know the causal relations behind the linear correlations is quite important.

The main difficulty found in this study stems from the lack of all relevant variables for analysis of QoL at sub-regional level (NUTS III), which leads us to work, in many cases, with proxy variables for those originally adopted.

ACKNOWLEDGMENTS

Andreia Dionísio and Maria Conceição Rego are pleased to acknowledge financial support from Fundação para a Ciência e a Tecnologia (grant UID/ECO/04007/2019).

REFERENCES

ADC, 2019. *Relatório do Desenvolvimento e Coesão.* (Coord.) Duarte Rodrigues and António Ramos. Agência de Desenvolvimento e Coesão, IP. [*Development and Cohesion Report.* (Coord.) Duarte Rodrigues and António Ramos. Development and Cohesion Agency, IP] Available at: https://www.adcoesao.pt/sites/default/files/relatorio_desenvolvimento_coesao_adc_2018.pdf (January, 2020).

Amado, C, Barreira, AP, Santos, S, and Guimarães, MH. 2019. Comparing the quality of life of cities that gained and lost population: An assessment with DEA and the Malmquist index. *Papers in Regional Science*. 98: 2075- 2097. https://doi.org/10.1111/pirs.12448.

Carta, M. G.; Aguglia, E.; Garcia, F.; Caraci, F.; Dell'Osso, L.; DiSciascios, G.; Drago, F.; Del Giudice, E.; Faravelli, C.; Hardoy, M. C.; Lecca, M. E.; Moro, M. F.; Calò, S.; Casacchia, M.; Angermeyer, M.; and Balestrieri, M. 2012. Quality of Life and Urban/Rural Living: Preliminary Results of a Community Survey in Italy, *Clinical Practice Epidemiology in Mental Health*,, 8, 169-174.

Dijkstra, L.; Poelman, H.; and Rodríguez-Pose, A. 2019. The geography of EU discontent, *Regional Studies*, https://doi.org/10.1080/00343404.2019.1654603.

Eurofound 2014. Quality of life in urban and rural Europe, European Foundation for the Improvement of Living and Working Conditions. *Publications Office of the European Union*, Luxembourg.

Ferrão, J. 2002. Portugal, Três Geografias em Recombinação(Portugal, recombining three geographies: spatialities, cognitive maps and territorial identities). *Lusotopie*, 2, 151-158.

Ferreira, P. and Dionísio, A. 2016. What are the conditions for good innovation results? A fuzzy-set approach for European Union. *Journal of Business Research*, 69(11), 5396-5400.

Fiss, P. 2011. Building better causal theories: a fuzzy set approach to typologies in organization research. *Academy of Management Journal*, 54(2), 393–420.

Hair, J. F., Black, W. C., Anderson, E. R., and Barry, J. B. 2014. *Multivariate Data Analysis*. 7th Ed, Prentice Hall.

OECD 2018. The continued importance of the "Beyond GDP" Agenda, in Beyond GDP: Measuring What Counts for Economic and Social Performance, *OECD Publishing*, Paris, https://doi.org/10.1787/9789264307292-4-en.

Paykani, T., Rafiey, H., and Sajjadi, H. 2018. A fuzzy set qualitative comparative analysis of 131 countries: which configuration of the structural conditions can explain health better?. *International Journal for Equity in Health*, 17(1), 10. doi:10.1186/s12939-018-0724-1.

Ragin, C. 2006. Set relations in social research: Evaluating their consistency and courage. *Political Analysis*, 14, 291–310.

Ragin, C. 1987. *The Comparative Method: Moving beyond Qualitative and Quantitative Strategies*. Berkeley: University of California Press.

Ragin, C. 2000. *Fuzzy-set social science*. Chicago: The University of Chicago Press.

Ragin, C. 2008. *Redesigning Social Inquiry: Fuzzy Sets and Beyond*. Chicago: The University of Chicago Press.

Ragin, C. and Fiss, P. 2008. Net effects versus configurations: An empirical demonstration. In C. Ragin (Ed.), *Redesigning social inquiry: Fuzzy sets and beyond* (pp. 190–212). Chicago: University of Chicago Press.

Rego, M. C., Freire, C., Ramos, I., Dionísio, A., Baltazar, M. S. and Lucas, M. R. 2017. Urban-Rural Connections and Development Perspectives in Portugal, *CEFAGE-UE Working Paper* 2017/04.

Rodrigues, C. F., Figueiras, R. and Junqueira, V. 2012. Desigualdade económica em Portugal (Economic Inequality in Portugal). *Fundação Francisco Manuel dos Santos*. Alêtheia Editores.

Rodríguez-Pose, A. 2018. The revenge of the places that don't matter (and what to do about it), *Cambridge Journal of Regions, Economy and Society*, 11 (1), 189–209, https://doi.org/10.1093/cjres/rsx024.

Shucksmith, M., Cameron, S., Merridew, T. and Pichler, F. 2009. Urban-rural differences in quality of life across the European Union. *Regional Studies*, 43(10), 1275-1289. https://doi.org/10.1080/003434008 02378750.

Tho, N. 2018. A Configurational Role of Human Capital Resources in the Quality of Work Life of Marketers: FsQCA and SEM Findings from Vietnam, Applied Research in Quality of Life, *Springer; International Society for Quality-of-Life Studies*, 13(2), 461-478.

Unidade de Missão para a Valorização do Interior, 2016. Programa Nacional para a Coesão territorial (PNCT) [*Mission Group for the Valorization of Interior, 2016. National Programme to Territorial Cohesion*]. Available at: https://www.portugal.gov.pt/pt/gc21/governo/programa/programa-nacional-para-a-coesao-territorial-/ficheiros-coesao-territorial/programa-nacional-para-a-coesao-territorial-pdf.aspx (accessed in January 2020).

Vis, B. 2012. The Comparative Advantages of fsQCA and Regression Analysis for Moderately Large-N Analyses. *Sociological Methods& Research*, 41(1), 168–198.

Vis, B., Woldendorp, J., Keman, H. 2007. Do miracles exist? Analyzing economic performance comparatively. *Journal of Business Research*, 60, 531–538.

Wagemann, C. and Schneider, C. 2010. Qualitative Comparative Analysis (QCA) and Fuzzy-sets: Agenda for a Research Approach and a Data Analysis Technique. *Comparative Sociology*, 9, 376-96.

Biographical Sketches

Andreia Dionísio

Affiliation: Universidade de Évora, Departamento de Gestão & CEFAGE,

Education: PhD in Management

Research and Professional Experience: Professor at the University of Évora specialized in statistics for business, data analysis, econometrics and econophysics and responsible for the Data Analysis courses. Her main areas of research have included the study of stock markets, global dependence, nonparametric methods and fuzzy set analysis. Over the last 15 years she has been involved in several research projects.

Publications from the Last 3 Years:

1. Ferreira, P. e Dionísio, A. (2019). "City Brand: what are the main conditions for better performance?", *Sustainability*, 11: 2-14. (doi.org/10.3390/su11143959)
2. Guedes, E., Ferreira, P., Dionísio, A. e Zebende, G. (2019). "An econophysics approach to study the effect of BREXIT referendum on European Union stock markets." *Physica A: Statistical Mechanics and its Applications,* 523: 1175-1182. (doi.org/ 10.1016/j.physa.2019.04.132)
3. Mothi, W; Dionísio, A.; Vieira, I. and Ferreira, P. (2019) "Regional and Global Integration of Asian Stock Markets". *Research in International Business and Finance*, 50(C): 357-368. (doi.org/10.1016/j.ribaf.2019.06.003)
4. Mothi, W; Dionísio, A.; Vieira, I. and Ferreira, P. (2019) "Financial contagion analysis in frontier markets: Evidence from the US subprime and the Eurozone debt crises." *Physica A: Statistical*

Mechanics and its Applications, 525: 1388-1398. (doi.org/10.1016/j.physa.2019.03.094)

5. Dias, R., Dionísio, A. e Silva, J.V. (2019) "Financial markets of the LAC region: does the crisis influence the financial integration?". *International Review of Financial Analysis*, 63: 160-173. (doi.org/10.1016/j.irfa.2019.02.008)
6. Mothi, W; Dionísio, A.; Ferreira, P. and Vieira, I. (2019) "Contagion of the Subprime Financial Crisis on Frontier Stock Markets: A Copula Analysis". *Economies, MDPI*, 7, 15. (doi.org/10.3390/economies7010015)
7. Ferreira, P. e Dionísio, A. (2019) "Using QCA to explain business demography in the European Union." *Journal of Business Research*, 101(C): 743-749. (doi.org/10.1016/j.jbusres.2018.12.074)
8. Mothi, W; Dionísio, A.; Ferreira, P. and Vieira, I. (2018) "Frontier markets' efficiency: mutual information and detrended fluctuation analyses." *Journal of Economic Interaction and Coordination,* 14: 551–572. (doi.org/10.1007/s11403-018-0224-9) (citations: Google Scholar: 1; SCOPUS: 1; Research Gate: 0)
9. Ferreira, P, Dionísio, A. e Correia, J. (2018). "Non-linear dependencies in African stock markets: Was subprime crisis an important factor?". *Physica A: Statistical Mechanics and its Applications*, 505, 680-687. (doi.org/10.1016/j.physa.2018.03.060) (citations: Google Scholar: 2; SCOPUS: 0; Research Gate: 3)
10. Morais, A., Fialho, A. e Dionísio A. (2018) "Is the accounting quality after the mandatory adoption of IFRS a random walk? Evidence from Europe." *Journal of Applied Accounting Research,* 19, 3: 334-350. (doi.org/10.1108/JAAR-07-2016-0073) (citations: Google Scholar: 5; SCOPUS: 0; Research Gate: 1)
11. Ferreira, P., Dionísio, A,. Guedes, E. e Zebende, G. (2018), "A sliding windows approach to analyse the evolution of bank shares in the European Union," *Physica A: Statistical Mechanics and its Applications*, 490: 1355-1367. (dx.doi.org/10.1016/j.physa.2017.08.095) (citations: Google Scholar: 9; SCOPUS: 8; Research Gate: 5)

12. Ferreira, P., Dionísio, A., Movahed, S. (2017). "Assessment of 48 Stock markets using adaptive multifractal approach." *Physica A: Statistical Mechanics and its Applications*, 486: 730-750. (dx.doi.org/ 10.1016/j.physa.2017.05.046) (citations: Google Scholar: 6; SCOPUS: 5; Research Gate: 2)
13. Ferreira, P., Nunes, J.M., Loures, L., Dionísio, A. (2017). "The behaviour of share returns of football clubs: An econophysics approach." *Physica A: Statistical Mechanics and its Applications*, 472: 136-144 (dx.doi.org/10.1016/j.physa.2017.01.022) (citations: Google Scholar: 1; SCOPUS: 1; Research Gate: 4)
14. Guedes, E., Dionísio, A., Ferreira, P. Zebende, G. (2017). "DCCA cross-correlation in blue-chips companies: A view of the 2008 financial crisis in the Eurozone." *Physica A: Statistical Mechanics and its Applications,* 479: 38-47. (dx.doi.org/10.1016/j.physa.2017. 02.065) (citations: Google Scholar: 20; SCOPUS: 17; Research Gate: 9)

Conceição Rego

Affiliation: Universidade de Évora, Departamento de Economia & CEFAGE,

Education: PhD in Economics

Research and Professional Experience: Professor of Economics on Department of Economics – University of Evora. As a researcher has coordinated and participated in several projects in the areas of Regional Development, Population Economics and Education Economics. Co-publisher of books, author and co-author of several articles published in books and journals, as well as national and international referee of publications and conferences.

Publications from the Last 3 Years:

1. Olo, Daniela; Correia, Leonida; Rego, M. Conceição (2020). The Main Challenges of Higher Education Institutions in the 21st Century: A Focus on Entrepreneurship, in (Eds) Ana Dias Daniel, Aurora A. C. Teixeira and Miguel Torres Preto, *Examining the Role of Entrepreneurial Universities in Regional Development*, IGI Global, 1-23. doi: 10.4018/978-1-7998-0174-0.
2. Rego, M. C.; Vieira, C.,Vieira, I. (2019). Educational Choices, Family Background and Social Mobility: Education and Social Mobility, in (Eds: Lusia Cagica Carvalho and Maria José Madeira) *Global Campaigning Initiatives for Socio-Economic Development*, Chapter 8, pp:128-146. doi: 10.4018/978-1-5225-7937-3
3. Lucas, M. R., Sousa, K. A., Ramos, I. J., & Rego, C. (2019). Desenvolvimento Sustentável, Economia Circular e Educação Empreendedora. In Gilson Pôrto Jr.(org.), *Pesquisa em inovação: múltiplos olhares rumo a uma convergência formativa* (recurso eletrônico):pp. 13-30. Palmas-TO:EDUFT.
4. http://download.uft.edu.br/?d=b7dc75c8-2b52-4259-a8c6-ed7c17e89907;1.0:Pesquisa%20em%20inova%C3%A7%C3%A3o%20multiplos%20olhares%20rumo%20a%20uma%20converg%C3%AAncia%20formativahttp://hdl.handle.net/10174/25855.
5. Dentinho, Tomaz; Damião, Pedro; Rego, Conceição (2018), Migration and regional development in Timor Leste in: *Population Change and Impacts in Asia and the Pacific,* Chapter: 12, Publisher: Springer, Editors: Jacques Poot, Matthiew Roskruge.
6. Rego, C., Lucas, M. R., Vieira, C., & Vieira, I. (2018). Entrepreneurial Ecosystems in low density regions: business incubation practices in Alentejo, in Luisa Cagica Carvalho (Ed.), *Handbook of Research on Entrepreneurial Ecosystems and Social Dynamics in a Globalized World*, Hershey, PA: IGI Global.
7. Rego, C., Ramos, I. J., Lucas, M. R., & Freire, M. (2018). Ensino Superior e a cooperação entre os países de língua portuguesa: onde estão as "salas de aula"?. In B. Cabrito, J. M. de Macedo, L. Cerdeira

(Orgs.), *Ensino Superior No Brasil E Em Portugal: atualidades, questões e inquietações* (pp. 149-169). Lisboa: EDUCA - Instituto de Educação da Universidade de Lisboa. [*Higher Education In Brazil And Portugal: news, questions and concerns*]

8. LUCAS, R., REGO, C., VIEIRA, C. e VIEIRA, I. (2017) Proximity and cooperation for innovative regional development: The case of the Science and Technology Park of Alentejo, in L. Carvalho (Ed.) *Handbook of Research on Entrepreneurial Development and Innovation within Smart Cities*, pp. 208-238, PA: IGI Global, Hershey, USA. doi: 10.4018/978-1-5225-1978-2.ch010.
9. Baltazar, M. S., Ramos, I. J., Rego, C., Dionísio, A., & Lucas, M. R. (2017). Jovens, *Migrações e Desenvolvimento Regional*. In Gilberta P. N. Rocha, Rolando L. Gonçalves, & Pilar D. de Medeiros (Org.). *Juventude(s): Pensar e Agir* (pp. 347-377). Ribeirão: Húmus. [*Youth (s): Think and Act*]
10. Lucas, M. R., Rego, C., Baltazar, M. S., Freire, M., Dionísio, A., & Ramos, I. J. (2017). Mobilidade Internacional e Escolhas dos Estudantes no Ensino Superior. O Programa Erasmus Em Portugal, *Revista FORGES*, vol. 5, nº 2, pp. 159-185. ISSN 2183-2722. [International Mobility and Student Choices in Higher Education. The Erasmus Program in Portugal, *FORGES Magazine*]

In: An Introduction to Fuzzy Sets
Editor: Valdemar F. Andersen

ISBN: 978-1-53618-012-1

Chapter 2

Economically Speaking, Are Happiness and HDI the Same? The Fuzzy-Set Approach

Andreia Dionísio[1] and Paulo Ferreira[2,*]

[1]Cefage, Department of Management, Universidade de Évora, Évora, Portugal

[2]VALORIZA - Research Center for Endogenous Resource Valorization, Instituto Politécnico de Portalegre, Portalegre, CEFAGE-UE, Universidade de Évora, Portugal

Abstract

The Human Development Index (HDI) has been an indicator used to evaluate various aspects of human development since the 1990s. Subject to several criticisms, namely the fact that it does not measure some kinds of relevant variables, the HDI has undergone several transformations in its calculation.

* Corresponding Author's E-mail: andreia@uevora.pt.

In parallel, other indices have been used to measure other situations. For example, the World Happiness Report (WHR) is an annual publication of the United Nations Sustainable Development Solutions Network, created to measure a happiness index, whose assumptions are different. The objective of this research is to identify the necessary and sufficient conditions for higher HDI levels and to determine whether the same pattern can be used to explain the happiness index.

The questions under study are: "Is the HDI mostly driven by economic forces? In the end, do happiness and human development have the same basis? Which are the Life and Economic necessary and sufficient conditions to reach high levels of HDI and WHI?"

Keywords: Human Development Index (HDI), World Happiness Index (WHI), fuzzy-set Qualitative Comparative Analysis (fsQCA), factor analysis

1. Introduction

The Human Development Index (HDI) is a composite index calculated yearly by the United Nations Development Programme (UNDP) and is used to measure countries' level of development, composed of three indices which measure three different situations: life expectancy, education and income. Created in the 1990s, HDI has undergone some changes and since 2010 the Human Development Report has calculated the Inequality-adjusted Human Development Index. Despite its general use, HDI comes in for some criticism, namely concerning the variables selected and the dimensions used in the respective calculation, the fact that it considers a high share of Gross Domestic Product (GDP), not considering distributive measures, the absence of ecological considerations, the lack of theoretical justification of the index or even data errors (see, for example, Sagar and Najam 1998, Herrero et al. 2010 or Wolff et al. 2011).

Previous to the creation of the HDI, the former king of Bhutan, Jigme Singye Wangchuck, stated in the 1970s that Happiness should be more important than GDP, creating the concept of Gross National Happiness (GNH), a philosophy that guides the government of Bhutan today. With the objective of attaining sustainable and equitable development, promoting and preserving culture and the environment, the use of GNH has also been subject to criticism (see, for example, Frame 2005 or Bates 2009).

More recently, the United Nations also started work on measuring happiness, presenting the World Happiness Report (WHR), which was first released in 2012. With a survey-based analysis, the WHR contains the World Happiness Index (WHI).

Some discussion is found on the comparison of happiness and development. For example, Blanchflower and Oswald (2005) identify a paradox between those dimensions in the particular case of Australia, later discussed by Leigh and Wolfers (2006). Some approaches measure national happiness indices (see, for example, Senasu et al. 2019).

In this paper, the objective is not to make any critical review of the HDI or the WHI, but rather, we propose to investigate whether the conditions to attain higher levels in those indices are the same, giving special attention to economic conditions. In fact, taking information of the HDI and the WHI for 2018, we correlated countries' positions in both indices and found a very strong correlation (0,8231), meaning than a higher position in the HDI implies a higher position in the WHI. So we will attempt to identify if both indices are affected by the same conditions.

Our main results point to smaller differences in necessary and sufficient conditions for HDI and WHI, and no primacy of economic conditions. We should note the importance of an elderly population as a necessary and core causal sufficient condition.

The remainder of the paper is organized as follows: Section 2 presents the data used for this study and methodology based on exploratory factor analysis and fuzzy-set qualitative comparative analysis, Section 3 presents the results obtained and Section 4 concludes.

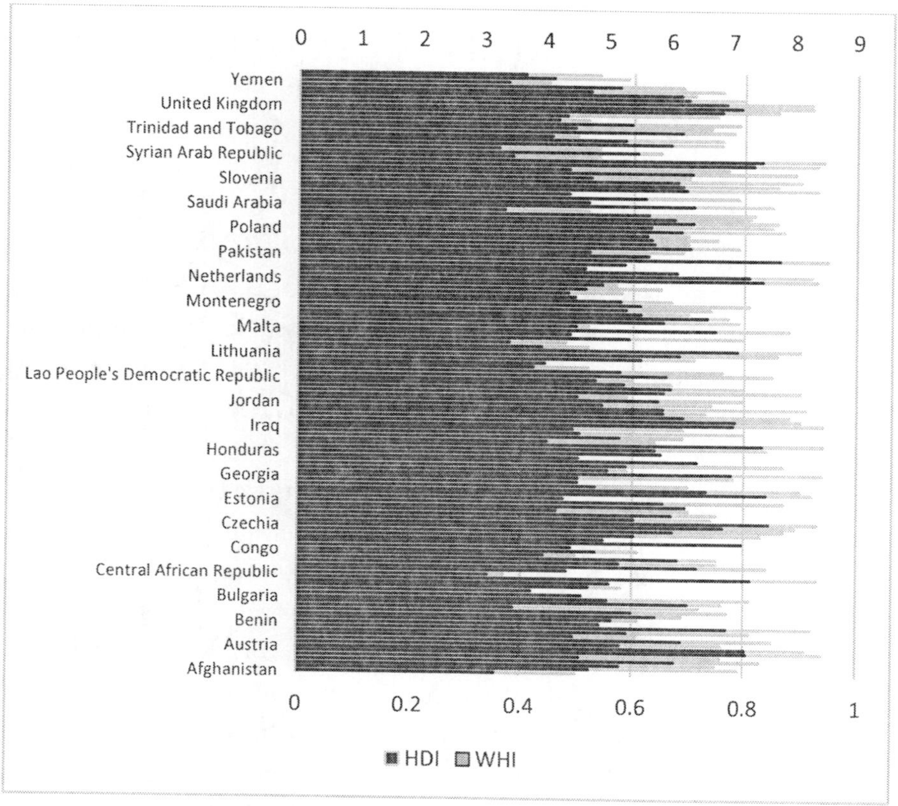

Figure 1. HDI and WHI 2018 for the sample of countries studied.

2. Data and Methods

2.1. The Data

This study intends to analyse and compare HDI and WHI as indexes based on well-being. The data related to HDI 2018 was obtained from the United Nations Development Programme (http://hdr.undp.org/en/data) and the WHI for 2018 was obtained in the World Happiness Report (http:// hdr.undp.org/en/data). Figure 1 shows the respective indexes for some of the countries studied.

From the whole set of variables with available information, we divided them in two main categories: Life Conditions (comprising information from the categories of Demography, Education, Environmental sustainability, Gender, Health, Human Security) and Economic Conditions (with information from Income/composition of resources, Mobility and communication, Socio-economic sustainability, Trade and financial flows, Work, employment and vulnerability). In appendix A, we present the variables used in each category and sub-category.

The division of variables in two categories is driven by the objectives of this research. The intention is to analyse and compare those two kinds of variables as necessary and sufficient conditions for HDI and discover if the situation is the same for WHI. Considering the availability of data, we retrieved data for 147 countries, as listed in Appendix B.

Given the enormous number of variables and the evidence of significant linear correlation between variables in the same sub-category, a factor analysis procedure was applied to the different sub-categories in order to reduce the number of independent variables to include in the fuzzy-set analysis. The general purpose of factor analytical techniques is to find a way to summarise the information contained in a number of original variables into a smaller set of new, composite dimensions (factors) with a minimum loss of information - that is, to search and define the fundamental dimensions assumed to underlie the original variables (Hair et al., 2014).

2.2. Methods

In order to achieve the objectives defined in this research, fuzzy-set Qualitative Comparative Analysis (fsQCA) was applied to the data obtained with factor analysis. This approach can determine the necessary and sufficient conditions for HDI and also WHI, establishing from the beginning a division of the variables studied: Life Conditions and Economic Conditions. The questions studied are: "Is the HDI mostly driven by economic forces? In the end, do happiness and human development have the same basis? What are the Life and Economic Conditions necessary and sufficient to reach high levels of HDI and WHI?"

This study uses fuzzy-set qualitative comparative analysis (fsQCA), a methodology which evaluates how conditions cause a specific outcome (Wagemann & Schneider 2010). While quantitative methodologies aim to find cause-effect relations between dependent and independent variables, this qualitative methodology (see for example, Vis (2012)) 'fits the causes-of-effects approach most because this approach aims to reveal the minimal (combinations of) conditions bringing about a particular outcome in specific cases.' Since the contribution of Ragin (1987), it has undergone several developments (Ragin, 2008). fsQCA was first used in social sciences such as sociology, but applications in economics and management are frequent, most of all, because of the potentialities of this methodology. For example, fsQCA can analyze different combinations of conditions in a given problem (Ragin, 2008), even when the dimension of the database (number of observations) is small or medium-sized (Vis, 2012). This feature gives it great potential, supporting numerous studies where the variables are usually qualitative and the number of observations does not assure the robustness of traditional statistical and econometric techniques.

QCA methodologies can capture both necessary and sufficient conditions. The consistency measure of Ragin (2006) measures the degree to which each case corresponds to a set-theoretic relationship given by a solution. It captures the proportion of cases which are consistent with the outcome and penalizes severe inconsistencies. For sufficient conditions, the consistency level is the measure used as the main background (Ragin, 2008).

Applying fsQCA implies calibration of the variables, given the importance of all variables lying between 0 and 1. The calibration process is fundamental to the methodology, and implies, according to Ragin (2000), 'substantive and theoretical knowledge relevant to set membership.' In the calibration process, the researcher is expected to identify the fully in set (the condition/outcome will have the value of 1), the fully out set (0 value), and also a crossover point (0.5) for all conditions and for the outcome. Given the lack of theoretical reasoning to define the bounds of the three thresholds for calibration, we base our calibration process on the percentile approach, which according to Ragin (2008), is appropriate when we have continuous data. The 'fully in' set was defined as the 95th percentile, the 'fully out' as the 5th percentile and the 'neither in nor out' point was defined by the median. We used the same criterion for all conditions and outcomes considered in this study. Results were estimated using the version of the fs/QCA software package (2.5).

3. RESULTS

This section starts by explaining the construction of orthogonal variables, using factor analysis. The variables (or factors) resulting from this analysis constitute the conditions used in analysis of the necessary and sufficient condition presented in the second sub-section.

3.1. Factor Analysis

As mentioned before, the original variables are grouped in the respective sub-categories and most of those are continuous. The scale was reduced and analysed using factor analysis (procedure Factor in SPSS version 22). The KMO index and the Bartlett's test for sphericity were used to confirm the significant correlation between the variables, indicating the quality of the factor analysis applied. The analysis employed principal components with Varimax rotation and extraction of the components was based on the eigenvalues criterion (Kaiser, 1974). The results are presented in Tables 1 to 7.

We emphasise in Table 1 variables with a coefficient greater than 0.6, making it possible to distinguish two components related to demography: The first component associated with the elderly population and the second component associated with the active population. A similar procedure was followed for the Education sub-category, where the results point to the existence of one component, which we call Education (see Table 2).

Table 1. Relationship between variables and different components. After rotation for Demography sub-category

Sub-category Demography	Component	
	Elderly Pop	Active Pop
Old-age (65 and older) dependency ratio (per 100 people aged 15-64)	0.969	0.188
Population aged 15-64 (millions)	0.139	0.946
Population aged 65 and older (millions)	0.955	0.214
Sex ratio at birth (male to female births)	0.106	0.603
Urban population (%)	0.462	0.525
Population under age 5 (millions)	-0.651	-0.712
Young age (0-14) dependency ratio (per 100 people aged 15-64)	-0.578	-0.789
KMO = 0.794		
Bartlett's Test = 1414.487**		

** 1% significance level.

Table 2. Relationship between variables and different components. After rotation for Education sub-category

Sub-category Education	Education
Expected years of schooling (years)	0.897
Expected years of schooling, female (years)	0.902
Expected years of schooling, male (years)	0.868
Mean years of schooling (years)	0.979
Mean years of schooling, female (years)	0.969
Mean years of schooling, male (years)	0.97
Population with at least some secondary education (% aged 25 and older)	0.954
Population with at least secondary educ, female (% aged 25 and older)	0.952
Population with at least some secondary educ, male (% aged 25 and older)	0.936
KMO = 0.779	
Bartlett's Test = 4548.385**	

** 1% significance level.

Table 3. Relationship between variables and different components. After rotation for Sustainability sub-category

Sub-category Sustainability	**Sustainability**
Mortality rate attributed to household and ambient air pollution (per 100,000 population)	0.904
Mortality rate attributed to unsafe water, sanitation and hygiene services (per 100,000 population)	0.938
Concentration index (exports) (value)	0.663
Rural population with access to electricity (%)	-0.917
KMO = 0.801	
Bartlett's Test = 397.74**	

** 1% significance level.

Table 4. Relationship between variables and different components. After rotation for Gender sub-category

Sub-category Gender	**Component**	
	GSEcon	GEmpl
Adolescent birth rate (births per 1,000 women aged 15-19)	-0.777	-0.016
Estimated gross national income per capita, female (2011 PPP $)	0.942	-0.057
Estimated gross national income per capita, male (2011 PPP $)	0.922	0.211
Share of employment in non-agriculture, female (% of total employment in non-agriculture)	0.187	-0.697
Total unemployment rate (female to male ratio)	0.114	0.937
Women with account at financial institution or with mobile money-service provider (% of female population aged 15 and older)	0.877	-0.224
Youth unemployment rate (female to male ratio)	0.055	0.838
KMO = 0.640		
Bartlett's Test = 827.12**		

** 1% significance level.

Table 5 shows the existence of two components. Given the relevant coefficients, we consider the first one as Mortality expectancy (Mort) and the second as Lack of immunization of infants (LImun). It is important to note that the second factor is based mostly on two variables, which may diminish the robustness of the respective factor.

Table 6 presents the rotated matrix for the Income and Employment sub-categories. The results show the existence of three components and given the coefficients with greatest weight, we called them Income_Emploment (IncEmp), Labour_Participation (LabPart) and Unemployment (Unemp).

Table 5. Relationship between variables and different components. After rotation for Health sub-category

Sub-category Health	Component	
	Mort.	LImun
Infants lacking immunization, DPT (% of one-year-olds)	0.657	0.705
Infants lacking immunization, measles (% of one-year-olds)	0.73	0.62
Life expectancy at birth (years)	-0.968	0.184
Life expectancy at birth, female (years)	-0.97	0.18
Life expectancy at birth, male (years)	-0.948	0.186
Mortality rate, infant (per 1,000 live births)	0.956	-0.136
Mortality rate, under-five (per 1,000 live births)	0.956	-0.092
Tuberculosis incidence (per 100,000 people)	0.684	-0.246
KMO = 0.743		
Bartlett's Test = 2943.35**		

** 1% significance level.

Table 6. Relationship between variables and different components. After rotation for Income and Employment sub-categories

Sub-categories Income and Employment	Component		
	IncEmp	LabPart	Unemp
GDP per capita (2011 PPP $)	0.917	0.097	-0.145
Gross national income (GNI) per capita (2011 PPP $)	0.919	0.1	-0.149
Employment in agriculture (% of total employment)	-0.885	0.264	-0.153
Employment in services (% of total employment)	0.877	-0.195	0.157
Employment to population ratio (% aged 15 and older)	-0.09	0.846	-0.495
Labour force participation rate (% aged 15 and older)	-0.119	0.944	-0.242
Labour force participation rate (% aged 15 and older), female	-0.111	0.932	-0.116
Unemployment, total (% of labour force)	-0.033	-0.218	0.959
Unemployment, youth (% ages 15-24)	0.07	-0.307	0.926
Vulnerable employment (% of total employment)	-0.883	0.26	-0.124
KMO = 0.705			
Bartlett's Test = 2649.39**			

** 1% significance level.

Table 7. Relationship between variables and different components. After rotation for Income and Employment sub-categories

Sub-categories of Mobility and Communication and Trade and Financial Flows	**Component**	
	Mob	Trade
Internet users, total (% of population)	0.876	0.153
Mobile phone subscriptions (per 100 people)	0.907	0.037
Exports and imports (% of GDP)	0.335	0.83
Foreign direct investment, net inflows (% of GDP)	-0.073	0.924
KMO = 0.547		
Bartlett's Test = 155.52**		

** 1% significance level.

Table 7 presents the components resulting from the sub-categories of Mobility and Communication, and Trade and Financial Flows. These two sub-categories were joined in order to improve the quality of the factor analysis, but even so, the KMO index shows the poor quality of this result. The two components obtained are respectively Mobility and Communication (Mob) and Trade.

The variable of Human Security is used in its original form, given the fact that this sub-category did not show a significant correlation with other variables.

3.2. Fuzzy Set Analysis

Regarding fsQCA, the first analysis assesses necessary conditions, with the results presented in Table 8. There, we present the necessary conditions both for higher levels of HDI and WHI as well as for lower levels (represented respectively by ~HDI and ~WHI). The relevant necessary conditions, those with consistency levels above 0.8, are similar for both indices. In fact, higher levels of HDI and WHI are affected positively by education, socioeconomic gender (GSEcon), income employment (IncEmp), mobility (Mobility) and elderly population (Elderly Pop) and negatively by sustainability, mortality (Mort) and human security

(HumanSec). The occurrence of lower levels of HDI and WHI is caused by higher levels of sustainability and mortality (Mort) and by the absence of education, socioeconomic gender (GSEcon), income employment (IncEmp) and mobility.

Table 8. Necessary conditions for HDI and WHI

	Outcome: HDI		Outcome: ~HDI		Outcome: WHI		Outcome: ~WHI	
Conditions	Consistency	Coverage	Consistency	Coverage	Consistency	Coverage	Consistency	Coverage
Elderly Pop	0.860	0.807	0.472	0.439	0.815	0.757	0.503	0.473
~Elderly Pop	0.401	0.434	0.792	0.849	0.432	0.463	0.741	0.802
Active Pop	0.769	0.786	0.533	0.541	0.754	0.764	0.552	0.565
~Active Pop	0.551	0.544	0.789	0.772	0.571	0.558	0.769	0.759
Education	0.917	0.886	0.462	0.442	0.835	0.799	0.513	0.496
~Education	0.423	0.442	0.881	0.913	0.473	0.490	0.792	0.829
Sustainability	0.426	0.452	0.861	0.905	0.448	0.471	0.806	0.855
~Sustainability	0.911	0.869	0.479	0.453	0.862	0.814	0.501	0.478
GSEcon	0.934	0.929	0.450	0.443	0.844	0.831	0.488	0.486
~GSEcon	0.440	0.446	0.594	0.565	0.478	0.480	0.830	0.843
Gempl	0.547	0.576	0.928	0.933	0.564	0.589	0.681	0.718
~Gempl	0.748	0.719	0.704	0.735	0.730	0.694	0.610	0.586
Mort	0.442	0.452	0.890	0.903	0.471	0.477	0.828	0.847
~Mort	0.905	0.893	0.460	0.450	0.849	0.830	0.489	0.483
Limun	0.573	0.634	0.628	0.688	0.606	0.663	0.619	0.684
~Limun	0.717	0.660	0.665	0.607	0.711	0.648	0.695	0.641
HumanSec	0.458	0.575	0.690	0.858	0.462	0.575	0.682	0.857
~HumanSec	0.887	0.743	0.659	0.546	0.885	0.734	0.662	0.554
IncEmp	0.912	0.936	0.433	0.440	0.847	0.861	0.458	0.470
~IncEmp	0.455	0.447	0.937	0.913	0.479	0.466	0.865	0.851
LabPart	0.629	0.631	0.683	0.679	0.661	0.657	0.672	0.675
~LabPart	0.680	0.684	0.629	0.627	0.673	0.670	0.658	0.662
Unemp	0.649	0.696	0.589	0.626	0.616	0.655	0.637	0.683
~Unemp	0.651	0.615	0.714	0.669	0.702	0.657	0.678	0.641
Mobility	0.894	0.877	0.471	0.458	0.840	0.816	0.510	0.500
~Mobility	0.448	0.461	0.874	0.891	0.486	0.495	0.812	0.836
Trade	0.623	0.699	0.603	0.670	0.606	0.674	0.619	0.695
~Trade	0.706	0.642	0.729	0.657	0.726	0.653	0.710	0.646

Table 9. Sufficient conditions to reach higher levels of HDI. Life conditions are represented on the left and economic conditions on the right

	HDI - life conditions			HDI - economic conditions				
	1	2	3	1	2	3	4	
Elderly Pop	•	•	•	●	●	●	●	GSEcon
Sustainability	⊗	⊗	⊗	⊗	•			GEmpl
Mort	⊗		⊗	•		•	•	IncEmp
Education		●	●	•	•	•		Mobility
LImun		⊗	⊗		⊗	⊗	⊗	LabPart
HumanSec			⊗		⊗		⊗	Unem
					•	•	•	Trade
RC	0.7714	0.5495	0.5702	0.6561	0.2797	0.4761	0.3772	RC
UC	0.1512	0.0123	0.0112	0.1051	0.0129	0.0129	0.0126	UC
Cons.	0.9594	0.9939	0.9976	0.9908	0.9787	0.9824	0.9904	Cons.
OSCov.	0.9274			0.8731				OSCov.
OSC	0.8789			0.8717				OSC

Legend: Cons: consistency; RC: Raw coverage; UC: Unique coverage; OSC: Overall Solution consistency; OSCov: Overall Solution coverage. ● Core causal condition (presence). • Peripheral causal condition (presence). ⊗ Core causal condition (absent). ⊗ Peripheral causal condition (absent).

The results are interesting since they show that issues such as sustainability and human security have an impact not in verifying higher values of the indices but help to lower those values.

We continue our analysis by considering the sufficient conditions. We follow the procedure proposed by Ragin and Fiss (2008), which shows the intermediate solutions but identifying core conditions (represented by larger symbols) and peripheral conditions (represented by smaller symbols). Tables 9 and 10 present the sufficient conditions for HDI and WHI respectively.

According to the results presented in Table 9, we identify three major combinations for life conditions and four combinations for economic conditions. The consistency of both overall solutions is above 0.85, overall solutions coverage being above 0.87. For the life conditions analysis, we must emphasize the combination of the elderly population and the absence of sustainability and mortality. This result may seem contradictory, but we must remember the construction of the sustainability condition. Ultimately, this is above all a lack of sustainability, an index of bad living conditions as regards air and water pollution and the availability of electricity in rural areas. The presence of education is also consensual, such as the absence or lack of immunization for children and human security. Concerning economic conditions, we should note the important presence of socioeconomic gender, income employment, mobility and trade. On the other hand, the absence of labour participation is not consensual and we believe this may be due to bad labour conditions in some countries, reducing the possibility of participating in labour decisions. The results are to some extent in line with those obtained in the necessary conditions analysis.

Table 10. Sufficient conditions to reach higher levels of WHI. Life conditions are on the left and economic conditions on the right

	WHI - life conditions			WHI - economic conditions					
	1	2	3	1	2	3	4	5	
Elderly Pop	●	●	●	•	⊗	•	•	•	GSEcon
Active Pop		⊗	●	⊗	⊗	•			GEmpl
Sustainability	⊗	•	⊗	•	⊗		•	•	IncEmp
Mort	⊗	•	⊗	•		•	•		Mobility
Education	•	⊗			⊗	•	•	⊗	LabPart
LImun	⊗	•			⊗	⊗		⊗	Unem
HumanSec						•	⊗	•	Trade
RC	0.6466	0.2266	0.5562	0.6195	0.2909	0.2806	0.4639	0.3683	RC
UC	0.1037	0.0172	0.0123	0.0912	0.0134	0.0131	0.0130	0.0101	UC
Cons.	0.9353	0.8044	0.9107	0.9266	0.8437	0.9725	0.9480	0.9577	Cons.
OSCov.	0.8738			0.8364					OSCov.
OSC	0.8013			0.8036					OSC

Legend: Cons: consistency; RC: Raw coverage; UC: Unique coverage; OSC: Overall Solution consistency; OSCov: Overall Solution coverage. ● Core causal condition (presence). • Peripheral causal condition (presence). ⊗ Core causal condition (absent). ⊗ Peripheral causal condition (absent).

The results presented in Table 10 show the sufficient conditions for WHI. In the life conditions group, we must emphasize three combinations, where the elderly population is the most relevant condition, with the absence of mortality and sustainability. For WHI, surprisingly, education does not have an important role as a sufficient condition, its absence being included in one of the combinations. Concerning economic conditions, we should note the presence of socioeconomic gender conditions, mobility and trade; and the absence of unemployment.

Globally, the results are quite similar to those of HDI, especially for economic conditions. The most important difference lies in education. In fact, education is a core causal condition for HDI, but not for WHI, which leads us to ask: is education not a basic condition for happiness?

Furthermore, our results do not show a primacy of economic conditions in HDI or WHI. In fact, probably the most important condition is the elderly population, which could be a sign of many other results: investment in health, in infrastructure/support measures for the elderly, for example. Basically, average life expectancy is the result of a great number of factors.

CONCLUSION

The literature contains several criticisms of the use of the HDI, some of them arguing that it fails to measure countries' true development, due to not considering possible relevant indicators, such as the population's happiness. In this paper, we analyse the main necessary and sufficient conditions to attain higher levels of HDI, based on an extended set of indicators provided by UNDP but not used in calculating the HDI. More than identifying the main conditions to achieve higher HDI values, this research analyses whether the conditions that affect HDI are the same as those affecting the WHI happiness index. We also analyse the importance of economic conditions compared to life conditions in those indexes.

Overall, our results point to a similar structure in terms of necessary and sufficient conditions for HDI and WHI and without primacy of economic conditions. It was curious to find the absence of Education as a sufficient condition for WHI, meaning that this component is probably not sufficiently valued by individuals and/or governments. In the end, the condition showing the greatest importance is elderly populations, leaving us to conclude that human development and happiness are based on longevity, sometimes, independently of the quality of that longevity.

REFERENCES

Bates, W. 2009. "Gross national happiness." *Asian Pacific Economic Literature* 23(2):1-16.

Blanchflower, D. and Oswald, A. 2005. "Happiness and the Human Development Index: The Paradox of Australia." *NBER Working Paper No. 11416*. Accessed on 5 January 2020, available at: https://www.nber.org/papers/w11416.

Eren, M., Çelik, A., and Kubat, A. 2014. "Determinants of the Levels of Development Based on the Human Development Index: A Comparison of Regression Models for Limited Dependent Variables." *Review of European Studies* 6(1):10-22.

Frame, B. 2005. "Bhutan: a review of its approach to sustainable development." *Development in Practice* 15(2):216-221.

Ferreira, P., and Dionísio, A. 2016. "What are the conditions for good innovation results? A fuzzy-set approach for European Union." *Journal of Business Research* 69(11):5396-5400.

Fiss, P. 2011. "Building better causal theories: a fuzzy set approach to typologies in organization research." *Academy of Management Journal* 54(2):393-420.

Hair, J. F., Black, W. C., Anderson, E. R., and Barry, J. B. 2014. *Multivariate Data Analysis*. 7th Ed, Prentice Hall.

Herrero, C., Martinez, R., and Villar, A. 2010. "Improving the Measurement of Human Development." *United Nations Development Programme Human Development Research Paper 2010/12*.

Leigh, A., and Wolfers, J. 2006. "Happiness and the Human Development Index: Australia is Not a Paradox." *The Australian Economic Review* 39(2):176-84.

Ragin, C. 2006. "Set relations in social research: Evaluating their consistency and courage." *Political Analysis* 14:291-310.

Ragin, C. 1987. *The Comparative Method: Moving beyond Qualitative and Quantitative Strategies*. Berkeley: University of California Press.

Ragin, C. 2000. *Fuzzy-set social science*. Chicago: The University of Chicago Press.

Ragin, C. 2008. *Redesigning Social Inquiry: Fuzzy Sets and Beyond*. Chicago: The University of Chicago Press.

Ragin, C., and Fiss, P. 2008. "Net effects versus configurations: An empirical demonstration." In: *Redesigning social inquiry: Fuzzy sets and beyond*, edited by C. Ragin, 190-212. Chicago: University of Chicago Press.

Senasu, K., Sakworawich, A., and Russ-Eft, D. F. 2019. "Developing Thai Happiness Index." *Social Indicators Research* 146(3):425-448.

Sagar, A., and Najamb, A. 1998. "The human development index: a critical review." *Ecological Economics* 25(3):249-264.

Vis, B. 2012. "The Comparative Advantages of fsQCA and Regression Analysis for Moderately Large-N Analyses." *Sociological Methods & Research* 41(1):168-198.

Vis, B., Woldendorp, J., and Keman, H. 2007. "Do miracles exist? Analyzing economic performance comparatively." *Journal of Business Research* 60:531-538.

Wagemann, C., and Schneider, C. 2010. "Qualitative Comparative Analysis (QCA) and Fuzzy-sets: Agenda for a Research Approach and a Data Analysis Technique." *Comparative Sociology* 9:376-96.

Wolff, H., Chong, H., and Auffhammer, M. 2011. *Classification, detection and consequences of data error: Evidence from the Human Development Index.* Accessed in January 2020, from Cornell University, School of Hospitality Administration http://scholarship. sha.cornell.edu/ articles/338.

Biographical Sketches

Andreia Dionísio

Affiliation: Universidade de Évora, Departamento de Gestão & CEFAGE.

Education: PhD in Management

Research and Professional Experience: Professor at the University of Évora specialized in statistics for business, data analysis, econometrics and econophysics and responsible for the Data Analysis courses. Her main areas of research have included the study of stock markets, global dependence, nonparametric methods and fuzzy set analysis. Over the last 15 years she has been involved in several research projects.

Publications from the Last 3 Years:

Dias, R., Dionísio, A. e Silva, J. V. (2019) "Financial markets of the LAC region: does the crisis influence the financial integration?" *International Review of Financial Analysis*, 63: 160-173 (doi.org/10.1016/j.irfa.2019.02.008).

Ferreira, P, Dionísio, A. e Correia, J. (2018). "Non-linear dependencies in African stock markets: Was subprime crisis an important factor?" *Physica A: Statistical Mechanics and its Applications*, 505, 680-687 (doi.org/10.1016/j.physa.2018.03.060) (citations: Google Scholar: 2; SCOPUS: 0; Research Gate: 3).

Ferreira, P. e Dionísio, A. (2019) "Using QCA to explain business demography in the European Union." *Journal of Business Research*, 101(C): 743-749 (doi.org/10.1016/j.jbusres.2018.12.074).

Ferreira, P. e Dionísio, A. (2019). "City Brand: what are the main conditions for better performance?" *Sustainability*, 11: 2-14 (doi.org/10.3390/su11143959).

Ferreira, P., Dionísio, A., Guedes, E. e Zebende, G. (2018), "A sliding windows approach to analyse the evolution of bank shares in the European Union," *Physica A: Statistical Mechanics and its Applications*, 490: 1355-1367 (dx.doi.org/10.1016/j.physa.2017.08.095) (citations: Google Scholar: 9; SCOPUS: 8; Research Gate: 5).

Ferreira, P., Dionísio, A., Movahed, S. (2017). "Assessment of 48 Stock markets using adaptive multifractal approach." *Physica A: Statistical Mechanics and its Applications*, 486: 730-750 (dx.doi.org/ 10.1016/j.physa.2017.05.046) (citations: Google Scholar: 6; SCOPUS: 5; Research Gate: 2).

Ferreira, P., Nunes, J. M., Loures, L., Dionísio, A. (2017). "The behaviour of share returns of football clubs: An econophysics approach." *Physica A: Statistical Mechanics and its Applications*, 472: 136-144 (dx.doi.org/10.1016/j.physa.2017.01.022) (citations: Google Scholar: 1; SCOPUS: 1; Research Gate: 4).

Guedes, E., Dionísio, A., Ferreira, P. Zebende, G. (2017). "DCCA cross-correlation in blue-chips companies: A view of the 2008 financial crisis in the Eurozone." *Physica A: Statistical Mechanics and its Applications*, 479: 38-47 (dx.doi.org/10.1016/j.physa.2017.02.065) (citations: Google Scholar: 20; SCOPUS: 17; Research Gate: 9).

Guedes, E., Ferreira, P., Dionísio, A. e Zebende, G. (2019). "An econophysics approach to study the effect of BREXIT referendum on European Union stock markets." *Physica A: Statistical Mechanics and its Applications*, 523: 1175-1182 (doi.org/10.1016/j.physa.2019.04.132).

Morais, A., Fialho, A. e Dionísio A. (2018) "Is the accounting quality after the mandatory adoption of IFRS a random walk? Evidence from Europe." *Journal of Applied Accounting Research*, 19, 3: 334-350 (doi.org/10.1108/JAAR-07-2016-0073) (citations: Google Scholar: 5; SCOPUS: 0; Research Gate: 1).

Mothi, W; Dionísio, A.; Ferreira, P. and Vieira, I. (2018) "Frontier markets' efficiency: mutual information and detrended fluctuation analyses." *Journal of Economic Interaction and Coordination*, 14: 551-572 (doi.org/10.1007/s11403-018-0224-9) (citations: Google Scholar: 1; SCOPUS: 1; Research Gate: 0).

Mothi, W; Dionísio, A.; Ferreira, P. and Vieira, I. (2019) "Contagion of the Subprime Financial Crisis on Frontier Stock Markets: A Copula Analysis." *Economies, MDPI*, 7, 15 (doi.org/10.3390/economies7010 015).

Mothi, W; Dionísio, A.; Vieira, I. and Ferreira, P. (2019) "Financial contagion analysis in frontier markets: Evidence from the US subprime and the Eurozone debt crises." *Physica A: Statistical Mechanics and its Applications*, 525: 1388-1398 (doi.org/10.1016/j.physa.2019.03.094).

Mothi, W; Dionísio, A.; Vieira, I. and Ferreira, P. (2019) "Regional and Global Integration of Asian Stock Markets." *Research in International Business and Finance*, 50(C): 357-368 (doi.org/10.1016/j.ribaf.2019. 06.003).

Paulo Ferreira

Affiliation: VALORIZA - Research Center for Endogenous Resource Valorization, Portalegre, Portugal; Instituto Politécnico de Portalegre, Portalegre, Portugal; CEFAGE-UE, IIFA, Universidade de Évora, Évora, Portugal.

Education: PhD in Management

Research and Professional Experience: Paulo Ferreira is an Economist and holds a PhD in Management. With several peer reviewed published papers in international reviews and technical books, his research is focused on the analysis of financial markets. Actually he is professor at the Polytechnic Institute of Portalegre, but with teaching experience in other higher education institutions, namely Universidade Moderna, Universidade Metodista (Angola), Instituto Superior de Línguas e Administração de Leiria, Universidade de Évora and Universidade Europeia. He is researcher at VALORIZA - Research Center for Endogenous Resource Valuation

(Portalegre) and in CEFAGE (University of Évora). He is currently Subdirector of the Agrarian Superior School of Elvas, of the Polytechnic Institute of Portalegre

Publications from the Last 3 Years:

Costa, N., Silva, C., Ferreira, P. (2019). "Long-range behaviour and correlation in cryptocurrencies DFA and DCCA analysis", *International Journal of Financial Studies*, 7: 51 (https://doi:10.3390/ijfs7030051) (citations: Google Scholar: 0; SCOPUS: 0; Research Gate: 0)

Ferreira, P. (2017). "Portuguese and Brazilian stock market integration: a non-linear and detrended approach", *Portuguese Economic Journal*, 16(1): 49-63 (doi:10.1007/s10258-017-0127-z) (citations: Google Scholar: 9; SCOPUS: 0; Research Gate: 7)

Ferreira, P. (2018). "Dynamic long-*range* dependences in the Swiss stock market", *Empirical Economics*, 1-33. (https://doi.org/10.1007/s00181-018-1549-x) (citations: Google Scholar: 1; SCOPUS: 0; Research Gate: 1)

Ferreira, P. (2018). "Efficiency or speculation? A time-varying analysis of European sovereign debt", *Physica A*, 490: 1295–1308. (https://doi.org/10.1016/j.physa.2017.08.137) (citations: Google Scholar: 3; SCOPUS: 5; Research Gate: 3)

Ferreira, P. (2018). "Long-range dependencies of Eastern European stock markets: A dynamic detrended analysis", *Physica A*, 505: 454–470. (https://doi.org/10.1016/j.physa.2018.03.088) (citations: Google Scholar: 7; SCOPUS: 7; Research Gate: 8)

Ferreira, P. (2018). "What detrended fluctuation analysis can tell us about NBA results", *Physica A*, 500: 92-96. (https://doi.org/10.1016/ j.physa.2018.02.050) (citations: Google Scholar: 2; SCOPUS: 1; Research Gate: 4)

Ferreira, P. (2018). "What guides Central and Eastern European stock markets? A view from detrended methodologies", *Post-Communist Economies*, 30(6): 805-819. (https://doi.org/10.1080/ 14631377.

2018.1460155) (citations: Google Scholar: 4; SCOPUS: 4; Research Gate: 3)

Ferreira, P. (2019). "Assessing the relationship between dependence and volume in stock markets: a dynamic analysis", *Physica A*, 516: 90-97. (https://doi.org/10.1016/j.physa.2018.09.187) (citations: Google Scholar: 2; SCOPUS: 1; Research Gate: 2)

Ferreira, P., Dionísio A., Correia, J. (2018). "Non-linear dependencies in African stock markets: Was subprime crisis an important factor?", *Physica A*, 505: 680-687 (https://doi.org/10.1016/j.physa.2018.03.060) (citations: Google Scholar: 4; SCOPUS: 3; Research Gate: 5)

Ferreira, P., Dionísio, A. (2017). "Long range dependence in G7 stock markets return rates using mutual information and detrended cross-correlation analysis". *Studies in Economics and Econometrics*, 41(1): 73 – 92 (citations: Google Scholar: 2; SCOPUS: 0; Research Gate: 1)

Ferreira, P., Dionísio, A. (2019). "City Brand: What Are the Main Conditions for Territorial Performance?", *Sustainability*, 11(14); 3959. (https://doi.org/10.3390/su11143959) (citations: Google Scholar: 0; SCOPUS: 0; Research Gate: 0)

Ferreira, P., Dionísio, A. (2019). "Using QCA to explain firm demography in the European Union", *Journal of Business Research*, 101: 743-749, (https://doi.org/10.1016/j.jbusres.2018.12.074) (citations: Google Scholar: 0; SCOPUS: 0; Research Gate: 0)

Ferreira, P., Dionísio, A., Guedes, E., Zebende, G. (2018). "A sliding windows approach to analyse the evolution of bank shares in the European Union", *Physica A*, 490: 1355-1367. (https://doi.org/10.1016/j.physa.2017.08.095) (citations: Google Scholar: 11; SCOPUS: 11; Research Gate: 8)

Ferreira, P., Kristoufek, L. (2017). "What is new about covered interest parity condition in the European Union? Evidence from fractal cross-correlation regressions", *Physica A*, 486: 554-566 (http://dx.doi.org/10.1016/j.physa.2017.05.085) (citations: Google Scholar: 13; SCOPUS: 11; Research Gate: 8)

Ferreira, P., Kristoufek, L. (2020). "Uncovered interest rate parity through the lens of fractal methods: Evidence from the European Union",

Physica A (https://doi.org/10.1016/j.physa.2020.124257) (citations: Google Scholar: 0; SCOPUS: 0; Research Gate: 0)

Ferreira, P., Kristoufek, L., Pereira, É. (2019). "DCCA and DMCA correlations of cryptocurrency markets", *Physica A* (https://doi.org/10.1016/j.physa.2019.123803) (citations: Google Scholar: 0; SCOPUS: 0; Research Gate: 0)

Ferreira, P., Loures, L. (2020). "An Econophysics study of the S&P Global Clean Energy Index", *Sustainability*, 12(2): 662 (https://doi.org/10.3390/su12020662) (citations: Google Scholar: 0; SCOPUS: 0; Research Gate: 0)

Ferreira, P., Loures, L., Nunes, J., Brito, P. (2018). "Are renewable energy stocks a possibility to diversify portfolios considering an environmentally friendly approach? The view of DCCA correlation coefficient", *Physica A*, 512: 675-681 (https://doi.org/10.1016/j.physa.2018.08.108) (citations: Google Scholar: 5; SCOPUS: 5; Research Gate: 4)

Ferreira, P., Loures, L., Nunes, J., Dionísio, A. (2017). "The behaviour of share returns of football clubs: An econophysics approach", *Physica A*, 472: 136-144 (http://dx.doi.org/10.1016/j.physa.2017.01.022) (citations: Google Scholar: 7; SCOPUS: 6; Research Gate: 7)

Ferreira, P., Pereira, É. (2019). "Contagion Effect in Cryptocurrency Market" *Journal of Risk and Financial Management*, 12: 115. (https://doi:10.3390/jrfm12030115) (citations: Google Scholar: 5; SCOPUS: 0; Research Gate: 1)

Ferreira, P., Pereira, E. (2019). "The impact of the Brexit referendum on British and European Union bank shares: a cross-correlation analysis with national indices." *Economics Bulletin*, 39(1): A35 (citations: Google Scholar: 1; SCOPUS: 0; Research Gate: 1).

Ferreira, P., Pereira, É., Pereira, H. (2020). "The exposure of European Union productive sectors to oil price changes", *Sustainability*, 12(4): 1620 (https://doi.org/10.3390/su12041620) (citations: Google Scholar: 0; SCOPUS: 0; Research Gate: 0)

Ferreira, P., Pereira, É., Silva, M. (2019). "The relationship between oil prices and the Brazilian stock market", *Physica A* (https://doi.org/

10.1016/j.physa.2019.123745) (citations: Google Scholar: 0; SCOPUS: 0; Research Gate: 0)

Ferreira, P., Pereira, E., Silva, M., Pereira, H. (2019). "Detrended correlation coefficients between oil and stock markets: the effect of the 2008 crisis", *Physica A*, 519: 86-96. (https://doi.org/10.1016/j.physa.2018.11.021) (citations: Google Scholar: 15; SCOPUS: 15; Research Gate: 5)

Ferreira, P., Silva, M., Santana, I. (2019) "Detrended Correlation Coefficients Between Exchange Rate (in Dollars) and Stock Markets in the World's Largest Economies", *Economies* 7(1): 9 (https://doi.org/10.3390/economies7010009) (citations: Google Scholar: 4; SCOPUS: 2; Research Gate: 2)

Filho, A., Pereira, E., Ferreira. P., Murari, T., Moret, M. (2018). "Cross-correlation analysis on Brazilian gasoline retail market", *Physica A*, 505: 550-557 (https://doi.org/10.1016/j.physa.2018.05.143) (citations: Google Scholar: 5; SCOPUS: 5; Research Gate: 4)

Guedes, E., Ferreira, P., Dionísio, A., Zebende, G. (2019). "An econophysics approach to study the effect of BREXIT referendum on European Union stock markets", *Physica A*, 523: 1175-1182. (https://doi.org/10.1016/j.physa.2019.04.132) (citations: Google Scholar: 1; SCOPUS: 1; Research Gate: 1)

Guedes, E., Zebende, P., Ferreira, P., Dionísio, A. (2017). "DCCA cross-correlation in blue-chips companies: a view of the 2008 financial crisis in the Eurozone", *Physica A*, 479:38-47. (http://dx.doi.org/10.1016/j.physa.2017.02.065) (citations: Google Scholar: 22; SCOPUS: 20; Research Gate: 13)

Kristoufek, L., Ferreira, P. (2018). "Capital asset pricing model in Portugal: Evidence from fractal regressions", *Portuguese Economic Journal* 17(3): 173-183 (https://doi.org/10.1007/s10258-018-0145-5) (citations: Google Scholar: 4; SCOPUS: 4; Research Gate: 4)

Loures, L., Ferreira, P. (2019). "Energy consumption as a condition for per capita carbon dioxide emission growth: The results of a qualitative comparative analysis in the European Union", *Renewable and Sustainable Energy Reviews*, 110: 220-225. (https://doi.org/10.1016/

j.rser.2019. 05.008) (citations: Google Scholar: 0; SCOPUS: 0; Research Gate: 0)

Mohti, W., Dionísio, A., Ferreira, P., Vieira, I. (2019) "Contagion of the Subprime Financial Crisis on Frontier Stock Markets: A Copula Analysis", *Economies* 7(1): 15 (https://doi.org/10.3390/economies 7010015) (citations: Google Scholar: 5; SCOPUS: 2; Research Gate: 3)

Mohti, W., Dionísio, A., Ferreira, P., Vieira, I. (2019). "Frontier markets' efficiency: mutual information and detrended fluctuation analyses", *Journal of Economic Interaction and Coordination*, 14(3): 551-572. (https://doi.org/10.1007/s11403-018-0224-9) (citations: Google Scholar: 1; SCOPUS: 1; Research Gate: 0)

Mohti, W., Dionísio, A., Vieira, I., Ferreira, P. (2019). "Financial contagion analysis in frontier markets: Evidence from the US subprime and the Eurozone debt crises", *Physica A*, 525: 1388-1398. (https://doi.org/10. 1016/j.physa.2019.03.094) (citations: Google Scholar: 2; SCOPUS: 2; Research Gate: 2)

Mohti, W., Dionísio, A., Vieira, I., Ferreira, P. (2019). "Regional and Global Integration of Asian Stock Markets", *Research in International Business and Finance*, 50: 357-368 (https://doi.org/10.1016/j.ribaf.2019.06.003) (citations: Google Scholar: 1; SCOPUS: 0; Research Gate: 0)

Movahed, S., Ferreira, P., Dionísio, A. (2017). "Assessment of 48 Stock markets using adaptive multifractal approach", *Physica A*, 486: 730–750 (https://doi.org/10.1016/j.physa.2017.05.046) (citations: Google Scholar: 9; SCOPUS: 6; Research Gate: 2)

Murari, T., Filho, A., Pereira, É, Ferreira, P., Pitombo, S., Pereira H., Santos, A., Moret, M. (2019). "Comparative analysis between hydrous ethanol and gasoline C pricing in Brazilian retail market", *Sustainability*, 11:4719 (https://doi:10.3390/su11174719) (citations: Google Scholar: 1; SCOPUS: 1; Research Gate: 1)

Pereira, É, Ferreira, P., Silva, M., Miranda, J., Pereira, H. (2019). "Multiscale network for 20 stock markets using DCCA", *Physica A*, 529: 121542. (https://doi.org/10.1016/j.physa.2019.121542) (citations: Google Scholar: 1; SCOPUS: 1; Research Gate: 1)

Pereira, É., Pereira, H., Carvalho, T., Ribeiro, L., Ferreira, P. (2019). "Policy in Brazil (2016-2018) threaten conservation of the Amazon rainforest", *Environmental Science and Policy*, 100: 8-12. (https://doi.org/10.1016/j.envsci.2019.06.001) (citations: Google Scholar: 11; SCOPUS: 3; Research Gate: 5)

Silva, L., Guedes, E., Ferreira, P., Dionísio, A., Zebende, G. (2019). "ρx,y between Open-Close stock markets", *Physica A*, 534: 122152. (https://doi.org/10.1016/j.physa.2019.122152) (citations: Google Scholar: 1; SCOPUS: 1; Research Gate: 0)

Tilfani, O., Ferreira, P., El Boukfaoui, Y. (2019). "Building multi-scale portfolios and efficient market frontiers using fractal regressions", *Physica A*, 532: 121758. (https://doi.org/10.1016/j.physa.2019.121758) (citations: Google Scholar: 0; SCOPUS: 0; Research Gate: 0)

Tilfani, O., Ferreira, P., El Boukfaoui, Y. (2019). "Dynamic cross-correlation and dynamic contagion of stock markets: a sliding windows approach with the DCCA correlation coefficient", *Empirical Economics* (forthcoming) (https://doi.org/10.1007/s00181-019-01806-1) (citations: Google Scholar: 1; SCOPUS: 0; Research Gate: 0)

Tilfani, O., Ferreira, P., El Boukfaoui, Y. (2019). "Revisiting stock market integration in Central and Eastern European stock markets with a dynamic analysis", *Post-Communist Economies*. (https://doi.org/10.1080/14631377.2019.1678099) (citations: Google Scholar: 0; SCOPUS: 0; Research Gate: 0)

Tilfani, O., Ferreira, P., El Boukfaoui, Y. (2020). "Multiscale optimal portfolios using CAPM fractal regression", *Post-Communist Economies*, 32(1): 77-112. (https://doi.org/10.1080/14631377.2019.1640983) (citations: Google Scholar: 0; SCOPUS: 0; Research Gate: 0)

In: An Introduction to Fuzzy Sets
Editor: Valdemar F. Andersen
ISBN: 978-1-53618-012-1

Chapter 3

IMPLEMENTATION OF FUZZY LOGIC AND NEURO-FUZZY IN INDUSTRY

Raheleh Jafari[1,*], Sina Razvarz[2], Alexander Gegov[3] and Saeed Razvarz[4]

[1]School of design, University of Leeds, Leeds, UK
[2]Departamento de Control Automatico, CINVESTAV-IPN (National Polytechnic Institute), Mexico City, Mexico
[3]School of Computing, University of Portsmouth, Buckingham Building, Portsmouth, UK
[4]Department of Mathematics, Tabriz branch, Islamic Azad University, Tabriz, Iran

ABSTRACT

Artificial intelligence, also named machine intelligence is the emulation of human intelligence procedures using machines, particularly computer systems. It is a branch of computer science, which purposes to generate intelligent machines and has become a fundamental part of the technology industry. Artificial intelligence makes it feasible for machines

* Corresponding Author's E-mail: r.jafari@leeds.ac.uk.

to acquire knowledge from experience, adapt to new inputs and accomplish human like tasks. In the last decades, artificial intelligence has had significant economic and societal benefits. Artificial intelligence methodologies such as fuzzy logic and neuro-fuzzy are broadly employed in various applications for example in business, marketing, control engendering, health care, and social services. In order to represent the applications of fuzzy set theory as well as neuro-fuzzy in industry, besides to provide a basis for future research, a literature review of these topics in industry is carried out in this book chapter. As some researchers have efficiently utilized fuzzy logic and neuro-fuzzy, indepth discussions are provided for stimulating future investigations.

Keywords: artificial intelligence, fuzzy set theory, industry

INTRODUCTION

Artificial intelligence and neuro-fuzzy are playing major roles in industrial en- gineering. Artificial neural networks are viewed as the most efficient method in recent decades that are largely employed in a vast range of implementations in various areas [1–5]. Artificial neural networks can be regarded as versatile tools, which have learning capability and model-free characteristics.

Fuzzy systems are proper tools for approximated reasoning, typically for the system having a mathematical design that is difficult to acquire [6–13]. Fuzzy set theory is regarded as the main problem designing and a solution technique. The primary contribution of fuzzy logic is its capability in presenting uncertain data. Fuzzy logic is employed to produce systems that are complicated to describe accurately. Currently, successful applications of fuzzy set theory in engineering have been reported. Industrial engineers encounter with different issues with inadequate and unsure information [14–20]. Fuzzy logic theory can be viewed as a proficient tool to resolve these kinds of issues. Fuzzy logic is considered as a useful approach for facilitating investigation in industrial engineering whenever the dynamics of the decision environment restrict the accurate evaluation of model parameters.

Neuro-fuzzy is employed in a vast extent of domains [21–38]. It is considered as a blend of artificial neural networks and fuzzy logic. Artificial neural networks can be used with fuzzy logic since it never makes a supposition on the probability distribution functions of data [39]. Normally, the neuro-fuzzy approach describes solutions superior when compared with artificial neural networks [40].

This book chapter indicates details of the applications of fuzzy set theory as well as neuro-fuzzy in industry. The most current researches in the area of fuzzy set theory and neuro-fuzzy are provided in this work. As some researchers have effectively used fuzzy logic and neuro-fuzzy, in-depth discussions are given for stimulating future studies. This work remaining sections are organized as follows. In Section 2 the usages of fuzzy logic in industrial engineering are presented. In Section 3 the usages of neuro-fuzzy in industrial engineering are presented. The conclusion of this book chapter is summarized in Section 4.

APPLICATIONS OF FUZZY LOGIC IN INDUSTRIAL ENGINEERING

Mathematical Preliminaries

Here some necessary definitions used in this paper are given.

Definition 1 (Fuzzy Variable)

If p is:

1) normal, there is $\zeta_0 \in R$ such that $p(\zeta_0) = 1$;
2) convex,$p[\lambda\zeta + (1-\lambda)\xi] \geq \min\{p(\zeta), p(\xi)\}, \forall \zeta, \xi \in R, \forall \lambda \in [0,1]$;
3) upper semi-continuous on R, $p(\zeta) \leq p(\zeta_0) + \varepsilon, \forall \zeta \in N(\zeta_0), \forall \zeta_0 \in R, \forall \varepsilon > 0, N(\zeta_0)$ is a neighborhood; or

4) $p^{+} = \{\zeta \in R, p(\zeta) > 0\}$ is compact, so P is a fuzzy variable, also the fuzzy set is expressed as E,$p \in E: R \rightarrow [0,1]$.

The fuzzy variable p is furthermore presented as below

$$p = A(\underline{p}, \overline{p}) \tag{1}$$

such that $\underline{p}$ is the lower-bound variable, $\overline{p}$ is the upper-bound variable, also A is taken to be a continuous function. The membership functions are applied for implicating the fuzzy variable p . The triangular membership function is defined as

$$p(\zeta) = F(a,b,c) = \begin{cases} \frac{\zeta-a}{b-a} & a \leq \zeta \leq b \\ \frac{c-\zeta}{c-b} & b \leq \zeta \leq c \\ 0 & otherwise \end{cases} \tag{2}$$

and the trapezoidal membership function is defined as

$$p(\zeta) = F(a,b,c,d) = \begin{cases} \frac{\zeta-a}{b-a} & a \leq \zeta \leq b \\ \frac{d-\zeta}{d-c} & c \leq \zeta \leq d \\ 1 & b \leq \zeta \leq c \\ 0 & otherwise \end{cases} \tag{3}$$

The fuzzy variable p, which has the dimension of ζ is dependent on the membership functions, such that (2) has three variables and (3) has four variables. For demonstrating the consistency of operations, the application initially lies within the $\alpha -$ level operation of the fuzzy number.

Definition 2 (Fuzzy Number)

A fuzzy number p related to the real value with $\alpha -$ level is defined as below

$$[p]^\alpha = \{a \in R : p(a) \geq \alpha\} \tag{4}$$

such that $0 < \alpha \leq 1, p \in E$.

If $p, \varphi \in E, \lambda \in R$, the fuzzy operations are as below:

Sum,

$$[p \oplus \varphi]^\alpha = [p]^\alpha + [\varphi]^\alpha + [\underline{p}^\alpha + \underline{\varphi}^\alpha, \overline{p}^\alpha + \overline{\varphi}^\alpha] \tag{5}$$

subtract,

$$[p \ominus \varphi]^\alpha = [p]^\alpha - [\varphi]^\alpha + [\underline{p}^\alpha - \underline{\varphi}^\alpha, \overline{p}^\alpha - \overline{\varphi}^\alpha] \tag{6}$$

or multiply,

$$\underline{\varpi}^\alpha \leq [p \odot \varphi]^\alpha \leq \overline{\varpi}^\alpha \text{or } [p \odot \varphi]^\alpha = A(\underline{\varpi}^\alpha, \overline{\varpi}^\alpha) \tag{7}$$

such that $\underline{\varpi}^\alpha = \underline{p}^\alpha \underline{\varphi}^1 + \underline{p}^1 \underline{\varphi}^\alpha - \underline{p}^1 \underline{\varphi}^1, \overline{\varpi}^\alpha = \overline{p}^\alpha \overline{\varphi}^1 + \overline{p}^1 \overline{\varphi}^\alpha - \overline{p}^1 \overline{\varphi}^1$, and $\alpha \in [0,1]$.

Hence, $[p]^0 = p^+ = \{\zeta \in R, p(\zeta) > 0\}$. As $\alpha \in [0,1]$, $[p]^\alpha$ is a bounded interval such that $\underline{p}^\alpha \leq [p]^\alpha \leq \overline{p}^\alpha$. The α −level of p between $\underline{p}^\alpha$ and $\overline{p}^\alpha$ is defined as

$$[p]^\alpha = A(\underline{p}^\alpha, \overline{p}^\alpha) \tag{8}$$

Let us take into consideration the below-mentioned unknown discrete-time nonlinear system

$$\overline{z}_{r+1} = \overline{h}(\overline{z}_r, v_r), w_r = \overline{q}(\overline{z}_r) \tag{9}$$

such that $v_r \in \mathfrak{R}^u$ is taken to be the input vector, $\overline{z}_r \in \mathfrak{R}^l$ is taken to be the internal state vector, also $w_r \in \mathfrak{R}^m$ is taken to be the output vector. $\overline{h}$ as well as $\overline{q}$ are considered as general nonlinear smooth functions $\overline{h}, \overline{q} \in C^\infty$.

Denoting $W_r = (w_{r+1}^T, w_r^T, \dots)^T$, $V_r = (v_{r+1}^T, v_r^T, \dots)^T$ If $\frac{\partial W}{\partial \overline{z}}$ is taken to be non-singular at $\overline{z} = 0, V = 0$ this results in the below-mentioned model

$$w_r = F(w_{r-1}^T, w_{r-2}^T, \dots, v_r^T, v_{r-1}^T, \dots) \tag{10}$$

Such that $F(\cdot)$ is taken to be an unknown nonlinear difference equation demonstrating the plant dynamics, v_r and w_r are taken to be the measurable scalar input and output, respectively. The nonlinear system (9) is stated as a NARMA model. The input of the nonlinear system can also be defined as follows

$$z_r = (w_{r-1}^T, w_{r-2}^T, \dots, v_r^T, v_{r-1}^T, \dots)^T \tag{11}$$

the output as w_r

Nonlinear systems defined in (9) can be demonstrated in the below-mentioned form,

$$w_r = \sum_{t=1}^{n} c_l h_l(z_r) \tag{12}$$

or

$$w_r + \sum_{l=1}^{m} d_l q_l(z_r) = \sum_{l=1}^{n} c_l h_l(z_r) \tag{13}$$

such that c_l as well as d_l are taken to be linear parameters, $h_l(z_r)$ as well as $q_l(z_r)$ are taken to be nonlinear functions. The variables related to these functions are calculable input and output. The popular example of this type of model can be the robot manipulator

$$A(s)\ddot{s} + B(s,\dot{s})\dot{s} + D\dot{s} + q(s) = \psi \tag{14}$$

(14) can be demonstrated as following

$$\sum_{l=1}^{n} W_l(s,\dot{s},\ddot{s})\vartheta_l = \psi \tag{15}$$

For identifying or control the systems (12), (13) or (15), the normal least square or adaptive techniques are used directly.

Here, we take into consideration the uncertain nonlinear systems, *i.e.*, the parameters c_l , d_l or ϑ_l are uncertain. The fuzzy nonlinear systems are modeled by linear-in-parameter models having uncertain parameters which are named fuzzy equations.

Remark

There exist different extensions of an equation to a fuzzy equation wherein the coefficients are taken to be fuzzy intervals [41]. For expanding an equation to a fuzzy equation, the interval definitions [42] can be applied to each α −cuts. In order to compute the membership function related to the set of solutions, the α −cut of the solution sets are stated using a conversion of the α −cuts of the fuzzy coefficients [43].

Two kinds of fuzzy equations have been used in below to model the fuzzy nonlinear system defined in (9),

$$w_r = c_1 h_1(z_r) \oplus c_2 h_2(z_r) \oplus \ldots \oplus c_n h_n(z_r) \tag{16}$$

or

$$\begin{aligned} c_1 h_1(z_r) \oplus c_2 h_2(z_r) \oplus \ldots \oplus c_n h_n(z_r) = \\ = d_1 q_1(z_r) \oplus d_2 q_2(z_r) \oplus \ldots \oplus d_m q_m(z_r) \oplus w_r \end{aligned} \tag{17}$$

As c_l and d_l are taken to be fuzzy numbers, hence the fuzzy operation $\oplus$ is utilized. (17) is the common form of (16), which is named as dual fuzzy equation.

In a particular instance, $h_l(z_r)$ has polynomial appearance,

$$c_1 z_r \oplus \ldots \oplus c_n z_r^n = d_1 z_r \oplus \ldots \oplus d_m z_r^m \oplus w_r \tag{18}$$

(18) is named dual fuzzy polynomial. If we utilize the dual fuzzy polynomial (18) for modeling the following nonlinear function

$$u_r = h(z_r) \tag{19}$$

hence the aim is to diminish error among the two outputs w_r and u_r. As w_r is taken to be a fuzzy number, also u_r is taken to be a crisp number, hence, the maximum of all points as the modeling error is defined as below

$$\max_{r}|w_r - u_r| = \max_{r}|w_r - h(u_{r)})| = \max_{r}|\varphi_r| \tag{20}$$

Such that $w_r = F(a(r), b(r), c(r))$, $\varphi_r = F(\varphi_1, \varphi_2, \varphi_3)$ that are stated in (2).

In [44], dual fuzzy equation (17) is applied for modeling the fuzzy nonlinear system (9). The procedure for controller design is based on obtaining v_r, in such a manner that the output related to the plant w_r follows the desired output w_r^*, or the trajectory tracking error is diminished

$$\min_{v_r}||w_r - w_r^*|| \tag{21}$$

The objective of control is to obtain a solution v_r for the below-mentioned dual fuzzy equation

$$\begin{aligned} c_1 h_1(z_r) \oplus c_2 h_2(z_r) \oplus \ldots \oplus c_n h_n(z_r) = \\ = d_1 q_1(z_r) \oplus d_2 q_2(z_r) \oplus \ldots \oplus d_m q_m(z_r) \oplus w_r^* \end{aligned} \tag{22}$$

such that $z_r = [w_{r-1}^T, w_{r-2}^T, \ldots, v_r^T, v_{r-1}^T, \ldots]^T$.

The fuzzy nonlinear system may furthermore be presented using partial differential equations as following

$$\frac{\partial^2 \phi(z,t)}{\partial t^2} + \frac{2}{t}\frac{\partial \phi(z,t)}{\partial t} = H(z, \phi(z,t), \frac{\partial \phi(z,t)}{\partial z}, \frac{\partial^2 \phi(z,t)}{\partial z^2}) \tag{23}$$

such that l as well as z are taken to be independent variables, ϕ is taken to be the dependent variable, *H* is considered as a nonlinear function of z ,ϕ , ϕ_z and ϕ_{zz} , furthermore the initial conditions for the partial differential equation (23) are stated as following:

$$\phi(z,0) = h(z), \phi_t(z,t) = q(z) \tag{24}$$

The below-mentioned fuzzy differential equation can be utilized for modeling the fuzzy nonlinear system (9),

$$\frac{d}{dt}z = h(z,v) \tag{25}$$

Such that z is taken to be the fuzzy variable, which correlates with the state z_r in (9) , $h(t,z)$ is taken to be a fuzzy vector function, which concerns with $h_1(z_r,v)$, and $\frac{d}{dt}z$ is taken to be the fuzzy derivative.

Definition 3 (Fuzzy Derivative)

The fuzzy derivative of h at z_0 is defined as follows

$$\frac{d}{dt}h(z_0) = \lim_{\tau \to 0} \frac{1}{\tau}[h(z_0+\tau) \ominus_{gH} h(z_0)] \tag{26}$$

such that $\ominus_{gH}$ is taken to be the Hukuhara difference [41], expressed as

$$z \ominus_{gH} w = c \Leftrightarrow \begin{cases} 1)\ z = w \oplus c \\ or\ 2)\ w = z \oplus (-1)c \end{cases} \tag{27}$$

The $\alpha -$ level related to the fuzzy derivative is defined as below

$$h(z,\alpha) = [\underline{h}(z,\alpha), \overline{h}(z,\alpha)] \tag{28}$$

such that $z \in E$ for every $\alpha \in [0,1]$.

By applying the $\alpha -$ level (aa) to $h(z,\alpha)$ in (afd) the following is concluded

$$[z \ominus_{gH} w]^{\alpha} = [\min\{\underline{z}^{\alpha} - \underline{w}^{\alpha}, \overline{z}^{\alpha} - \overline{w}^{\alpha}\},$$
$$ma\,x\{\underline{z}^{\alpha} - \underline{w}^{\alpha}, \overline{z}^{\alpha} - \overline{w}^{\alpha}\}] \tag{29}$$

hence, two functions are acquired: $\underline{h}[v,\underline{z}(\phi,\alpha),\overline{z}(\phi,\alpha)]$ and $\overline{h}[v,\underline{z}(\phi,\alpha),\overline{z}(\phi,\alpha)]$ Therefore,, the fuzzy differential equation (25) is equal with the below-mentioned four ordinary differential equations

$$\begin{cases} \frac{d}{dt}\underline{Z}(\alpha) = \underline{h}[v,\underline{z}(\phi,\alpha),\overline{z}(\phi,\alpha)] \\ \frac{d}{dt}\overline{Z}(\alpha) = \overline{h}[v,\underline{z}(\phi,\alpha),\underline{z}(\phi,\alpha)] \end{cases}$$

$$or \begin{cases} \frac{d}{dt}\underline{Z}(\alpha) = \overline{h}[v,\underline{z}(\phi,\alpha),\underline{z}(\phi,\alpha)] \\ \frac{d}{dt}\overline{Z}(\alpha) = \underline{h}[v,\underline{z}(\phi,\alpha),\overline{z}(\phi,\alpha)] \end{cases} \quad (30)$$

Fuzzy If-Then Rules

Fuzzy logic usually applies If-Then rules. A single fuzzy If-Then rule is stated as follows:

If p is x_1 Then q is y_2 such that x_1 and y_2 are taken to be the linguistic variables stated by fuzzy sets on the domains P and Q , respectively. The If sector of the rule " p is x_1 " is named as the antecedent and the Then sector of the rule" q is y_2 " is named as the consequent.

Furthermore, the conditional statement is defined as follows:

$$If\ x_1 then\ y_2\ or\ x_1 \rightarrow y_2$$

Example 1. Speed as well as pressure of a steam engine is defined with the below linguistic expression If speed is slow then pressure is high The statement is demonstrated in Figure 1.

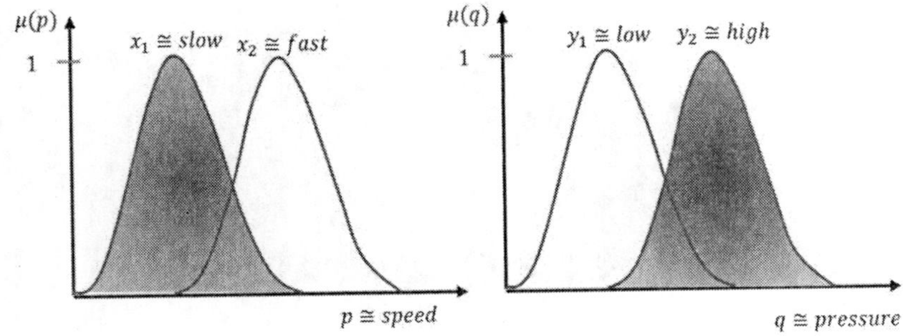

Figure 1. Fuzzy If-Then rule.

Background and Related Work

Broad researches have been done on implementing fuzzy set theory to research problems in industrial engineering.

In [46] fuzzy models on the basis of present worth are suggested for measuring manufacturing flexibility. The fuzzy models on the basis of present worth are mainly engineering economics decision designs such that the fuzzy cash flows, as well as discount amounts, are determined in the form of triangular fuzzy numbers. In order to create the model, fuzzy present worth formulas related to the manufacturing flexibility components are generated.

In [47] a fuzzy set technique for a multi-criteria choice of object-oriented simulation software to analyze the production system is proposed. The technique applies fuzzy set theory as well as algebraic operations of fuzzy numbers for characterizing simulation software, hence the ability and disability of every alternative is compared.

In [48] the location of a robot in a universal map from only sonar-based information is recognized. Fuzzy set theory is used for modeling sonar data also fuzzy triangulation is applied to recognize the position as well as the orientation of the robot. A fuzzy location area is achieved such that every point in the area contains a grade of confidence of being the real location of the robot.

Figure 2. Steel laminated gear.

Figure 3. Cast iron gear.

In [49] the knowledge related to the production volume as well as loading conditions using fuzzy sets are developed. The technique for the construction of gears, containing the illustration of the software outcomes as well as their explanation is provided. In order to construct a reliable lot of gears steel laminated gear (see Figure 2) or cast iron gear (See Figure 3) are utilized.

In [50] a model for examining the safety of engineering systems is suggested such that combines the fuzzy sets theory with apparent logic. The security of the system is presented by individually examining the treatment of its elements with regard to failure probability, result severity as well as failure result likelihood. It has been mentioned that the coalition of the fuzzy set theory in risk evaluation patterns is an effective approach for dealing with vague problems [51].

Fuzzy Expert System

The fuzzy expert system applies fuzzy logic in lieu of Boolean logic. It is the most general utilization of fuzzy logic. It is utilized in various areas such as:

- Linear and nonlinear control
- Pattern recognition
- Financial systems

The fuzzy inference systems for industrial position factor analysis is partitioned into three major parts: the fuzzifier, the knowledge management and the defuzzifier (Figure 4). The fuzzifier and the defuzzifier transform exterior information in fuzzy values and conversely. The knowledge management applies the fuzzy rules related to the knowledge basis which permits for estimate logic [52].

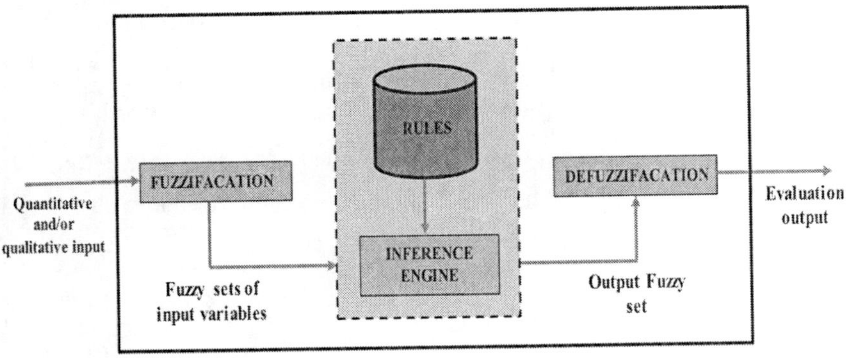

Figure 4. Fuzzy expert sysyem.

Advantageous and Drawbacks of Fuzzy Based Systems

Fuzzy based systems have the following advantageous:

1) Excellent accuracy on results.
2) Same as human logic.
3) Utilizations of easy mathematical models to solve real world problems linear or non-linear.
4) Utilized for fast operations as well as decision control.
5) Extremely effective for rule based modeling as well as membership evaluations.
6) Fuzzy based systems have the following drawbacks:
7) Poor speed and lengthy running time needed.
8) Absence of real-time answer.
9) For accurate outcomes require to involve an important quantity of data, which furthermore rises rules for reasoning.

Example of Fuzzy Logic Technique

The water storage cistern has two inlet valves M_1, M_2, and two exhaust valves N_3, N_4. The areas related to the valves are

$A_1 = F(0.016,0.018,0.019), A_2 = F(0.011,0.026,0.041), A_3 = F(0.005,0.007,0.009), A_4 = F(0.043,0.052,0.061))$. The velocities regarding to the flow are $p_1 = \left(\frac{v}{5}\right) e^{v}, p_2 = v cos(v), p_3 = \cos\left(\frac{v}{4}\right), p_4 = \frac{v}{6}$ The aim is to obtain the control variable , such that $q = (5.354,8.621,41.772)$. The mass balance related to the water storage cistern is defined as follows [51],

$$\varphi A_1 p_1 \oplus \varphi A_2 p_2 = \varphi A_3 p_3 \oplus \varphi A_4 p_4 \oplus q \quad (31)$$

such that φ is the density of the water. The exact solution is $v_0 = 3$. In order to estimate solution, three techniques like Steepest descent technique, Adomian decomposition technique, and Ranking technique can be used. The errors related to these techniques are given in Figure 5.

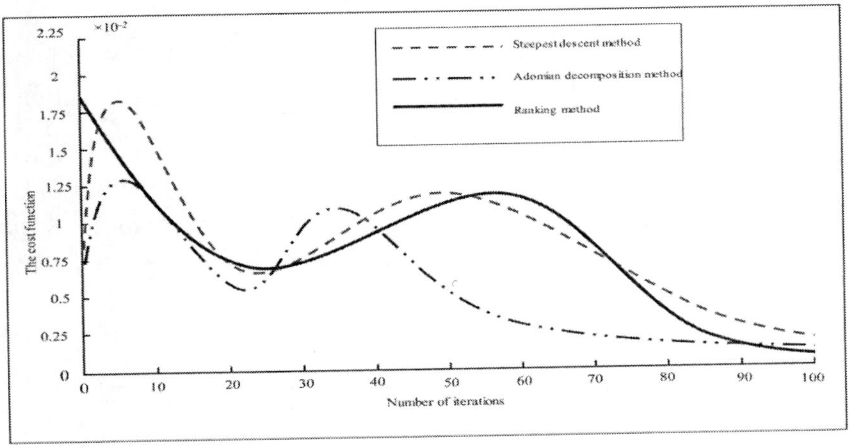

Figure 5. Estimated errors.

Applications of Neuro-Fuzzy in Industrial Engineering

Neural Systems

Biological Neural Network

The brain is made of nearly 100 billion neurons, such that every neuron has 1000–10000 connections with other neurons. They are in touch via electrical signals. Neurons composed of a cell body that have a nucleus, which controls the activity of the cell. Every neuron contains a body with branches named dendrites that capture signals from other neurons via a type of thread pipeline called axon and transfer information into the cell, see Figure 6. Neurons are formed in an entirely linked network and treat as a messenger in capturing and dispatching impulses. The consequence is an intelligent brain, which is able to learn, predict and recognize.

Artificial Neural Networks

In 1940, McCulloch-Pitts suggested the initial artificial neuron model that is later utilized in different feed-forward artificial neural networks like multilayer perceptrons [53]. The artificial neuron carries out a linear conversion via a weighted addition using the scalar weights. According to Figure 7, the activation function is computed by obtaining the addition of the input and its corresponding weight $(\tilde{y}_l * \widetilde{w}_l)$ for every input unit. The bias amount, $\tilde{b}$ can be added to the addition outcome, $\tilde{s}$. Afterward, the new suits into the activation function $f(\tilde{S})$. Finally, the outcome of the activation function is applied for predicting the output neuron J .

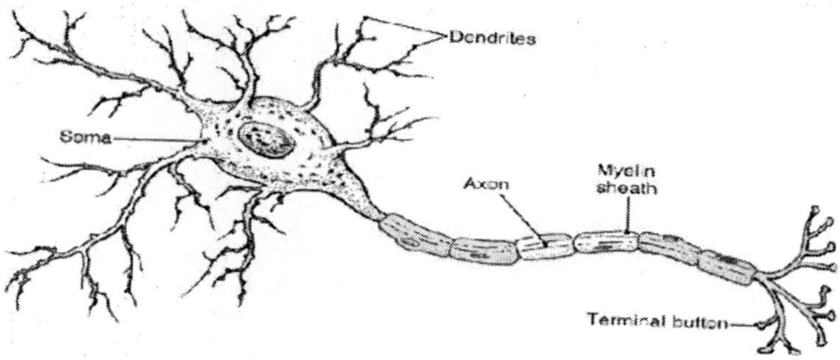

Figure 6. Biological neural network.

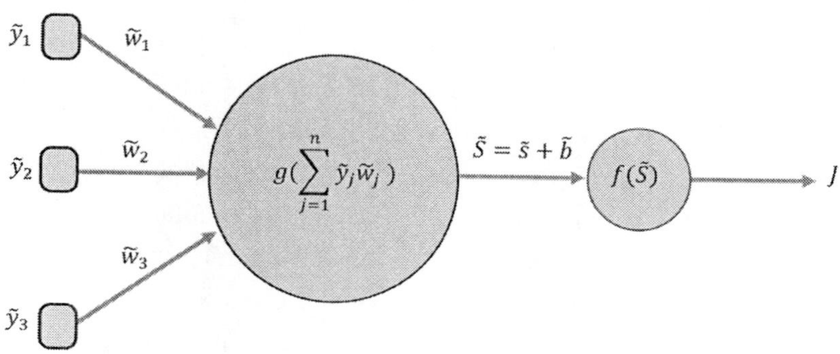

Figure 7. Structure of an artificial neuron.

There exist different types of activation functions like Sigmoid, Tanh, and ReLU, which are shown in Figures 8, 9, and 10, respectively.

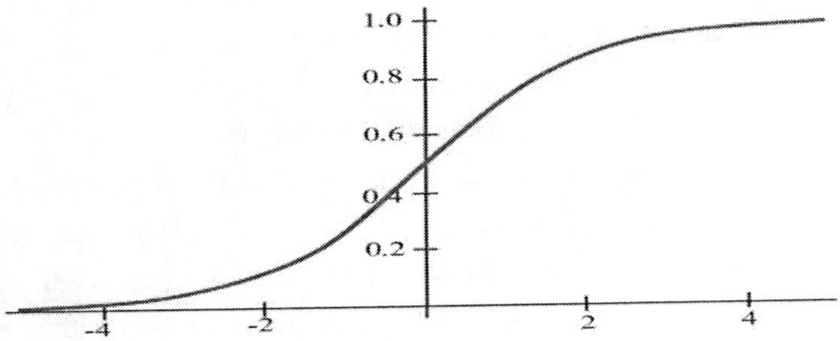

Figure 8. Sigmoid activation function.

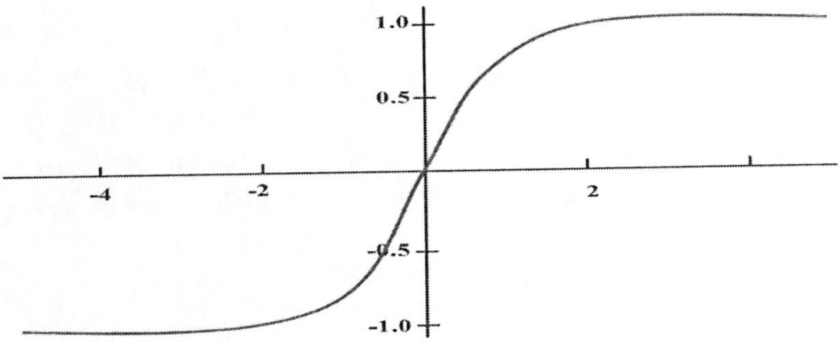

Figure 9. Tanh activation function.

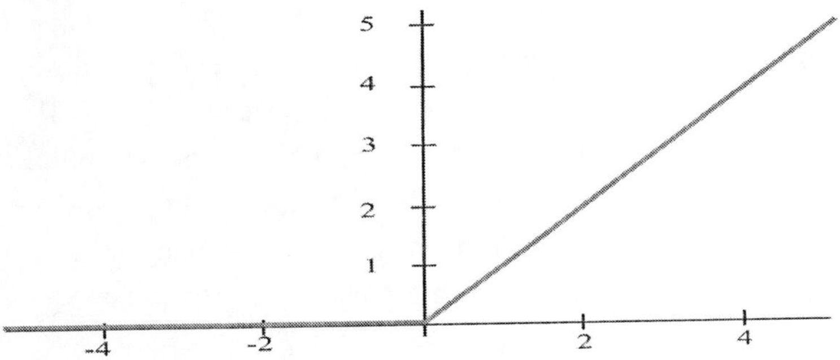

Figure 10. ReLU activation.

The process, which the neurons are linked with each other contains an important effect on the function of the artificial neural network. Similar to actual neurons, artificial neurons obtain either excitatory or inhibitory inputs. Excitatory inputs make the adding device of the subsequent neuron to sum whilst the inhibitory inputs make it to minus.

Feedback is considered as another kind of linkage such that the output related to a layer has a feedback loop to the input of a prior layer, or even to the equal layer. Two kinds of structures can be stated recognized based on the inexistence and existence of feedback linkage in a network. The feedforward structure does not contain any linkage back from the output to the input neurons, hence it won't have any recording of its former output amounts, see Figure 11.

Feedback structure contains linkage from output to input neurons. Every neuron contains one extra weight like an input, which permits an extra grade of freedom while attempting to decrease the training error, see Figure 12. Feedback neural network retains a memory of former state, therefore subsequent state apart from relying on input signals relies on the former states of the network.

BACKGROUND AND RELATED WORK

In [54], a system to diagnose the fault in AC motors using soft computing is proposed. The kernel related to the system is taken to be a neuro-fuzzy system called the fuzzy adaptive system ART-based, which allows the diagnosis of the fault in a case that it is in progress. The system has been examined on an AC motor when 15 nondestructive fault kinds are produced, obtaining top quality of detection as well as classification. The information reserved in the neuro-fuzzy system is obtained using a fuzzy rule set with a desirable rate of interpretability, also minus incoherency between the obtained rules.

In [55] the usage of the hybrid systematic scheme in multi-objective market issues is investigated. The main problem is proposed like an unstructured real-world problem in such a manner that the aims may not be illustrated mathematically, also just a set of historical data is used. Traditional techniques and likewise meta-heuristic techniques are broken in such situations. Therefore, a systematic scheme utilizing the hybrid of intelligent systems, especially the fuzzy rule base, as well as neural networks, may lead the decision-maker into noninferior solutions.

In [56] a diagnostic technique to optimize the energy performance as well as to maximize the operational time using artificial neural network and fuzzy logic is proposed. The basic modules, as well as the corresponding variables required to assess the module "health state," are recognized. Furthermore, the resulting improvement of the monitoring system is illustrated.

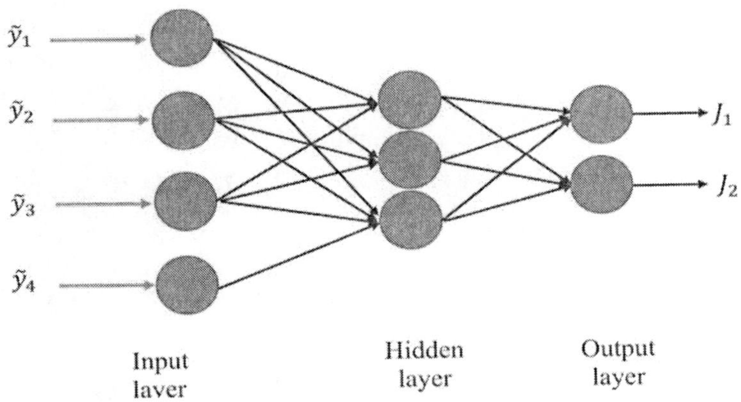

Figure 11. Feedforward neural network.

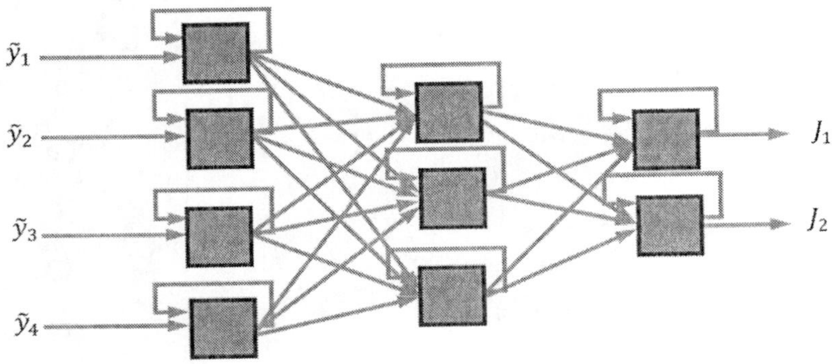

Figure 12. Feedback neural network.

In [57] a neural as well as a fuzzy virtual flight recorder is built, which records aircraft control surface deviations. The neural network simulator interacts with the recorders. The results of these two re-builders are taken to be the control surface deviations, which reduce the efficiency index on the basis of the variances among the accessible data and the result of the simulator.

In [58] fuzzy techniques are utilized for consecutive driving of long-domain autonomous global micro-rovers that need increasing of the domain. A perfect variety of methods such as fuzzy-based control, real-time logic, also fast as well as robust rover location approximation on the basis of odometry, angular measure sensing, and effective stereo observation are applied.

In [59] neuro-fuzzy transition control is proposed for a vehicle with variable loads. Automobile loads, as well as the driver's intent, are approximated by fuzzy inference, also optimal gear-shift moment choice is obtained using neural network technique. Convenient driving along with variable loads is obtained.

Example of Neuro-Fuzzy Technique

The deformation related to a solid cylindrical rod is based on the stiffness S, the given force to it f, the locations related to the forces P, also the diameter related to the rod b. The locations of the forces are $P_1 = F(0.4,0.7,0.8,0.9), P_2 = F(0.9,0.11,0.12,0.13), P_3 = F\,(0.8, 0.9, 0.14, 0.15)$. The area related to the rod is $R = \frac{\pi}{4}d^2$. The exterior forces are $f_1 = v^5, f_2 = \sqrt{v}, f_3 = e^v$. The aim is to obtain the amount of the control force v, such that $j = (0.000761, 0.000992, 0.001345, 0.001657)$. The tension relation is defined as follows [60],

$$\frac{P_1 f_1}{RE} \oplus \frac{P_2(f_1+f_2)}{RE} = \frac{P_3 f_3}{RE} \oplus j \tag{32}$$

such that $d = 0.02$, $E = 70 \times 10^9$. The exact solution is $v_0 = 5$. In order to estimate solution, Neural network technique can be applied. The error related to this technique is given in Figure 13.

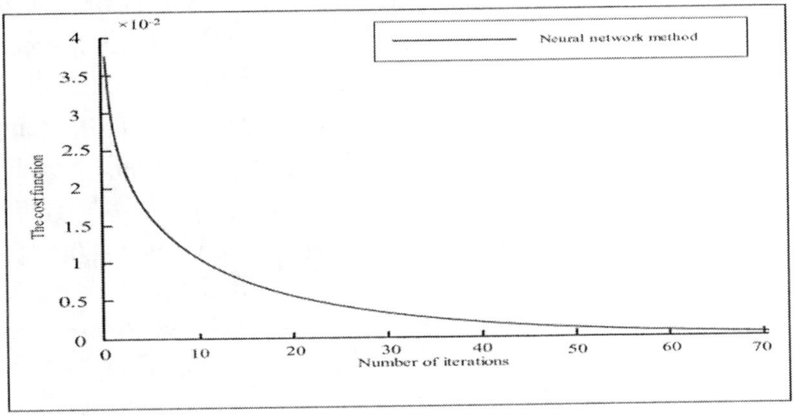

Figure 13. Estimated error.

CONCLUSION

In this book chapter, the usages of fuzzy logic and neuro-fuzzy in industrial engineering are presented. Fuzzy logic and neuro-fuzzy are considered as two important computational intelligence approaches. employments of these approaches could be taken into consideration as a cheap, highly impressive, as well as more reliable alternative tools. Hence, fuzzy logic and neuro-fuzzy techniques can provide more potency to problem resolving in comparison with other approaches.

REFERENCES

[1] Graupe, D.: Chapter 1–12. In: Chen W, Mlynski DA (eds) *Principles of artificial neural networks, Advanced Series in circuits and systems,* 3, 1st edn. World Scientific, p. 4e189, (1997).

[2] Parrales, A., Colorado, D., Diaz-Gomez, J. A., Huicochea, A., Alvarez, A., Hernandez J. A.: New void fraction equations for two-phase flow in helical heat exchangers using artificial neural networks, *Appl. Therm. Eng.,* 130, 149-160, (2018).

[3] Alimoradi, H., Shams, M.: Optimization of subcooled flow boiling in a vertical pipe by using artificial neural network and multi objective genetic algorithm, *Appl. Therm. Eng.,* 111, 1039-1051, (2017).

[4] Chang, W., Chu, X., Fareed, A. F. B. S., Pandey, S., Luo, J., Weigand, B., Laurien, E.: Heat transfer prediction of supercritical water with artificial neural networks, *Appl. Therm. Eng.,* 131, 815-824, (2018).

[5] Mohanty, S.: Estimation of vapour liquid equilibria for the system carbon dioxide–difluoromethane using artificial neural networks, *International Journal of Refrigeration,* 29, 243–249, (2006).

[6] Negoita, C. V., Ralescu, D. A.: *Applications of Fuzzy Sets to Systems Analysis,* Wiley, New York, (1975).

[7] Zadeh, L. A.: Probability measures of fuzzy events, *Journal of Mathematical Analysis and Applications,* 23, 421-427, (1968).

[8] Zadeh, L. A.: Calculus of fuzzy restrictions, *Fuzzy sets and Their Applications to Cognitiveand Decision Processes, Academic Press, New York, 1975*, 1-39.

[9] Zadeh, L. A.: Fuzzy logic and the calculi of fuzzy rules and fuzzy graphs, *Multiple-Valued Logic,* 1, 1-38, (1996).

[10] Jiang, W., Xie, C., Luo, Y., Tang, Y.: Ranking Z-numbers with an improved ranking method for generalized fuzzy numbers, *Journal of Intelligent and Fuzzy Systems,* 32, 1931-1943, (2017).

[11] Yaakob, A. M., Gegov, A.: Fuzzy rule based approach with Z-numbers for selection of alternatives using TOPSIS, *Proceedings of the IEEE International Conference on Fuzzy Systems,* 1-8, (2015).

[12] Zamri, N., Ahmad, F., Rose, A. N. M., Makhtar, M.: A Fuzzy TOPSIS with Z-Numbers Approach for Evaluation on Accident at the Construction Site, *Proceedings of the International Conference on Soft Computing and Data Mining,* 41-50, (2016).

[13] Wakami, N., Araki, S., Nomura, H.: Recent applications of fuzzy logic to home appliances, *in Proc. IEEE Int. Conf. Industrial Electronics, Control, and Instrumentation, Maui, HI,* 155–160, (1993).

[14] Aliev, R. A., Alizadeh, A. V., Huseynov, O. H.: The arithmetic of discrete Z-numbers, *Inform Sci,* 290, 134-155, https://doi.org/ 10.1016/j.ins.2014.08.024, (2015).

[15] Aliev, R. A., Huseynov, O. H., Zeinalova, L. M.: The arithmetic of continuous Z-numbers, *Inform Sci,* 373, 441-460, https://doi.org/ 10.1016/j.ins.2016.08. 078, (2016).

[16] Jiang, W., Xie, C., Zhuang, M., Shou, Y., Tang, Y.: Sensor Data Fusion with Z-Numbers and Its Application in Fault Diagnosis, *Sensors (Basel),* 16, doi:10.3390/s16091509, (2016).

[17] Nuriyev, A. M.: Application of Z-Numbers Based Approach to Project Risks Assessment, *European Journal of Interdisciplinary Studies,* 5, 67-73, ISSN 2411-4138.

[18] Abiyev, R. H., Uyar, K., Ilhan, U., Imanov, E., Abiyeva, E.: Estimation of Food Security Risk Level Using Z-Number-Based Fuzzy System, *Journal of Food Quality,* 2018, Article ID 2760907, 9 pages, https://doi.org/10.1155/2018/2760907, (2018).

[19] Babanlia, M. B., Huseynov, V. M.: Z-number-based alloy selection problem, *12th International Conference on Application of Fuzzy Systems and Soft Computing, ICAFS 2016, 29-30 August 2016, Vienna, Austria, Procedia Computer Science* 102, 183-189, (2016).

[20] Kim, J. H., Kim, K. S., Sim, M. S., Han, K. H., Ko, B. S., An application of fuzzy logic to control the refrigerant distribution for the multi type air conditioner, *in Proc. IEEE Int. Fuzzy Systems Conf.,* 3, 1350-1354, (1999).

[21] Jafari, R., Razvarz, S., Gegov, A.: Neural Network Approach to Solving Fuzzy Nonlinear Equations using Z-Numbers, *IEEE Transactions on Fuzzy Systems,* doi: 10.1109/TFUZZ.2019.2940919, (2019).

[22] Jafari, R., Yu, W., Li, X., Razvarz, S.: Numerical solution of fuzzy differential equations with Z-numbers using bernstein neural

networks, *International Journal of Computational Intelligence Systems,* 10, 1226-1237, (2017).

[23] Werbos, P. J.: Neuro-control and elastic fuzzy logic: Capabilities, concepts, and applications, *IEEE Trans. Ind. Electron.,* 40, 170–180, (1993).

[24] Dote, Y., Hoft, R. G.: Intelligent Control—Power Electronics Systems, *Oxford, U.K.: Oxford Univ. Press,* (1998).

[25] Sato, S., Arai, Y., Hirota, K.: Pattern recognition using fuzzy inference with lacked input data, *in Proc. IEEE Int. Conf. Fuzzy Systems, San Antonio, TX,* 100–104, (2000).

[26] Lee, C. H., Teng, C. H.: Identification and control of dynamic systems using recurrent fuzzy neural networks, *IEEE Trans. Fuzzy Syst.,* 8, 349-366, (2000).

[27] Patra, S. K., Mulgrew, B.: Fuzzy implementation of a Bayesian equaliser in the presence of inter-symbol and co-channel interference, *Inst. Elect. Eng. Proc. Commun.,* 145, 323-330, (1998).

[28] Choi, B. J., Han, S. W., Hong, S. K.: Refrigerator temperature control using fuzzy logic and neural network, *in Proc. IEEE Int. Symp. Industrial Electronics, Pretoria, South Africa,* 186-191, (1998).

[29] Liu, C. W., Su, M. C., Tsay, S. S., Wang, Y. J.: Application of a novel fuzzy neural network to real-time transient stability swings prediction based on synchronized phasor measurements, *IEEE Trans. Power Syst.,* 14, 685–692, (1999).

[30] Lin, F. J., Wai, R. J., Lin, C. H., Liu, D. C.: Decoupled stator-flux-oriented induction motor drive with fuzzy neural network uncertainty observer, *IEEE Trans. Ind. Electron.,* 47, 356-367, (2000).

[31] Akcayol, M. A.: Application of adaptive neuro-fuzzy controller for SRM, *Advances in Engineering Software,* 35, 129-137, (2004).

[32] Jang, J.: ANFIS: Adaptive network-based fuzzy inference systems, *IEEE Transactions on Systems Man and Cybernetics,* 23, 665, (1993).

[33] Stathacopoulou, R., Magoulas, G. D., Grigoriadou, M., Samarakou, M.: Neurofuzzy knowledge processing in intelligent learning environments for improved student diagnosis, *Information Sciences,* 170, 273-307, (2005).

[34] Sevarac, Z.: Neuro fuzzy reasoner for student modeling, *IEEE International Conference on Advanced Learning Technologies,* 740–744, (2006).

[35] Jafari, R., Razvarz, S., Gegov, A., Paul, S.: Fuzzy Modeling for Uncertain Nonlinear Systems Using Fuzzy Equations and Z-Numbers. Contributions *Presented at the 18th UK Workshop on Computational Intelligence*, September 5-7, 2018, Nottingham, UK, (2018).

[36] Razvarz, S., Jafari, R., Gegov, A.: A New Computational Method for Solving Fully Fuzzy Nonlinear Systems, *10th International Conference,* ICCCI 2018, Bristol, UK, September 5-7, 2018, Proceedings, Part I. (2018).

[37] Jafari, R., Yu, W.: Fuzzy Modeling for Uncertainty Nonlinear Systems with Fuzzy Equations. In: *Mathematical problems in Engineering*, Vol. 2017, Article ID 8594738, 10 pages. (2017) https://doi .org/10.1155/2017/8594738.

[38] Yu, W., Jafari, R.: Fuzzy Modeling and Control with Fuzzy Equations and Z-Number. In: *IEEE Press Series on Systems Science and Engineering,* Wiley-IEEE Press, ISBN-13: 978-1119491552, 208 Pages. (2019).

[39] Castellano, G., Fanelli, A. M.: Variable selection using neural-network models, *Neurocomputing,* 31, 1-13, (2000).

[40] Babuska, R., Verbruggen, H.: Neuro-fuzzy methods for nonlinear system identification, *Annual Reviews in Control,* 27, 73–85, (2003).

[41] Lodwick, W., Dubois., D.: Interval linear systems as a necessary step in fuzzy linear systems, *Fuzzy Sets Syst,* 281, 227-251, (2015).

[42] Shary, S.: Solving the tolerance problem for interval linear equations, *Interval Comput.,* 2, 4-22, (1994).

[43] Nguyen, H.: A note on the extension principle for fuzzy sets, *J. Math. Anal. Appl.* 64, 369-380, (1978).

[44] Jafari, R., Yu, W., Li, X.: Fuzzy differential equations for nonlinear system modeling with bernstein neural networks, *IEEE Access,* 4, 9428-94367, (2016).

[45] Bede, B., Stefanini, L.: Generalized differentiability of fuzzy-valued functions, *Fuzzy Sets Syst.,* 230, 119-141, (2013).

[46] Kahraman, C., Beskese, A., Ruan, D.: Measuring flexibility of computer integrated manufacturing systems using fuzzy cash flow analysis, *Information Sciences,* 168, 77-94, (2004).

[47] Cochran, J. K., Chen, H. N.: Fuzzy multi-criteria selection of objectoriented simulation software for production system analysis, *Computers Operations Research,* 32, 153-168, (2005).

[48] Demirli, K., Türksen, I. B.: Sonar based mobile robot localization by using fuzzy triangulation, *Robotics and Autonomous Systems,* 33, 109-123, (2000).

[49] Cristian-Gheorghe, T., Felicia, B., George, B.: Fuzzy logic applications in gears manufacturing, *Advanced Materials Research,* 1036, 1028-1032, (2014).

[50] Wang, J., Yang, J. B., Sen, P.: Safety analysis and synthesis using fuzzy sets and evidential reasoning, *Reliability Engineering and System Safety,* 47, 103-118, (1995).

[51] Streeter, V. L., Wylie, E. B., Benjamin, E.: Fluid mechanics, *4th Ed, Mc. Graw-Hill Book Company,* (1999).

[52] Czogala, E., Leski, J.: Fuzzy and Neuro-Fuzzy Intelligent Systems, ser, *Studies in Fuzziness and Soft Computing, Springer,* 47, 2000.

[53] Warren, S. M., Walter, P.: A logical calculus of the ideas immanent in nervous activity, *Bull. Math. Biophys.,* 5, 115–133, (1943). http://www.cse.chalmers.se/ coquand/AUTOMATA/mcp.pdf.

[54] Palmero, S. G. I., Santamaria, J. J., de la Torre, E. J. M., Gonzalez, J. R. P.: Fault detection and fuzzy rule extraction in AC motors by a neuro-fuzzy ART-based system, *Engineering Applications of Artificial Intelligence,* 18, 867-874, (2005).

[55] Gholamian, M. R., Ghomi, S. M. T. F., Ghazanfari, M.: A hybrid systematic design for multiobjective market problems: a case study in crude oil markets, *Engineering Applications of Artificial Intelligence,* 18, 495-509, (2005).

[56] Barelli, L., Bidini, G.: Design of the measurements validation procedure and the expert system architecture for a cogeneration internal combustion engine, *Applied Thermal Engineering,* 25, 2698-2714, (2005).

[57] Napolitano, M. R., Casanova, J. L., Windon, D. A., Seanor, B., Martinelli, D.: Neural and fuzzy reconstructors for the virtual flight data recorder, *IEEE Trans. Aerosp. Electron. Syst.,* 35, 61-71, (1999).

[58] Martin-Alvarez, A., Volpe, R., Hayati, S., Petras, R.: Fuzzy reactive piloting for continuous driving of long range autonomous planetary micro-rovers, *in Proc. IEEE Aerospace Conf.,* 2, 127–135, (1999).

[59] Hayashi, K., Shimizu, Y., Dote, Y., Takayama, A., Hirako, A.: Neuro fuzzy transmission control for automobile with variable loads, *IEEE Trans. Contr. Syst. Technol.,* 3, 49-53, (1995).

[60] Beer, F. P., Johnston, E. R.: Mechanics of materials, *2nd Edition, Mcgraw- Hill.* (1992).

In: An Introduction to Fuzzy Sets
Editor: Valdemar F. Andersen
ISBN: 978-1-53618-012-1

Chapter 4

LOTFI ZADEH'S THEORY OF FUZZY SETS IN DECISION-MAKING PROCESS FOR OIL AND GAS PRODUCTION

Vidadi Bakhishov[1,*] and Elmira Isayeva[2]
[1]Agency of Small Business Development, Ministry of Economics of Azerbaijan Republic, Baku, Azerbaijan
[2]Institute of Physics of Azerbaijan Academy of Sciences, Baku Azerbaijan

ANNOTATION

The oil field and oil production belong to the class of the big, complex and hierarchical systems. Therefore, there is a need for good modeling. Today there are many means of modeling. But they all can consider only one concrete type of uncertainty. Using the L. Zadeh theory of fuzzy sets we can consider all types of uncertainties at the same time and to make a decision of what model is the best in such fuzzy environment. As result we

* Corresponding Author's Email address: vidadi.bah@gmail.com.

can get up the more intellectually control and monitoring of system of oil and gas production.

Often many scientists call Zadeh the father of fuzzy logic. What does the “fuzzy” logic mean and what is its difference from “usual” logic? The usual logic is the Aristotle’s logic that dominates in science and demands either to be or not to be of things in world, third for them is except. However, Zadeh have worked out a new mathematical theory called theory fuzzy sets (FST). If the well-known theory of probabilities considers only “probability” measure then Zadeh has to deal with new levels of uncertainties such as “possibility” and “necessary”. For this aim, for his theory he introduces “member function” that becomes very original in mathematics. Thus, Zadeh's FST can be used by us to recognize and predict the new levels of uncertainties, for example, in the oil and gas production.

It is known in the oil and gas deposit there are many kinds of uncertainties. Here the developed quantitative methods of decision-making (such as the theory of games, the minimax theory, maximizing of expected usefulness, methods of maximum likelihood, the analysis “expenses - efficiency” and others) help to choose the best of a set of possible decisions only in the conditions of one concrete type of uncertainty. All these methods are based on the probability theory therefore here the decision-making process is impossible. The probability theory deals with the randomness, but the main source of uncertainty is the fuzziness. The Zadeh FST considers this kind of uncertainties and allows to use the decision-making process. In this case we will be able to use decision-making process in a new way and and it is possible the intellectual control and monitoring of system of oil and gas production. In our project we will use known in the oil production the variant modeling, stochastic modeling, evolution and integration of models. We will take models of the different classes, different types of difficulties, different levels of the description of the object, different means and technologies of their construction, interpretation and application. At multiple stochastic modeling, there are a representative ensemble (set) of models of oil and gas deposit. The choice of the best variant of model occurs in the fuzzy environment. The statement “the chosen model is the best” is inexact. Such inaccuracy is expressed by a fuzzy set of all good models in which it is impossible to specify strict border between the elements which are belonging and not belonging to it. Therefore, there is a membership function of x to a fuzzy set A [1]. How will go decision-making process in fuzzy environment is considered by L. Zadeh [2] and we will apply its results to a case of oil and gas deposit. Thus X = {x} are the set of all versions of models. Then fuzzy set of the best models A in X is the set of pairs: $A = \{x, \mu_A(x)\}, \quad x \in X$, where $\mu_A(x)$ is membership function x to A .The problem of estimation of membership functions $\mu_A(x)$ is the main in fuzzy sets theory and there are different methods. For example, to construct membership functions basing on the information of the expert. Also, we will use the known in FST the fuzzy clustering, the base fuzzy algorithm of α – cut of average. After finding set of pairs $\{x_1, \mu_A(x_1);\ x_2, \mu_A(x_2);....\}$ we should decide another problem - problem of abstraction which plays central role in recognition of forms in FST. Let's assume, that $\mu_A(x)$ are the known membership functions for all $x \in X$. The main elements of decision-making process in FST are: a) set of alternatives X, in our case X is the set of variants of models x, b) set of the restrictions G, with which must be accounted when choosing the necessary alternative. In our case the restrictions are the geological data about petroleum deposit – the scale, depth, quantity, quality and etc -and other data – the communications, urbanizations, system of pipelines and etc. c) a function of preference. Joint influence of fuzzy restrictions G can be presented by the crossing of $G_1 \cap G_2 \cap \ldots \cap G_n$. A membership function will be:

$$\mu_{G_1 \cap G_2 \cap ..G_n} = \mu_{G_1}(x) \wedge \mu_{G_2}(x) \wedge \ldots \wedge \mu_{Gn}(x) = \min(\mu_{G_1}(x), \ldots, \mu_{Gn}(x))$$

Optimal decision is the precise subset D^M of the fuzzy set D determined as:

$$\mu_{D^M}(x) = \begin{cases} \max \mu_D(x), & x \in K \\ 0 \quad , & x \notin K \end{cases}$$

where K is the set of such X for which $\mu_D = \max$, every X from D^M is the maximizing decision.

During formation of models there will be their various state (variants of models) X_t, where t = 0,1 …. Here the entrance signals U_t (in our case - above mentioned restrictions G), $U_t \in U = \{\alpha_1, \alpha_2 \ldots\}$. U = 1, 2 take place. It is clear that the state X_{t+1} depends from X_t and U_t and it is described by the equation of evolution: $X_{t+1} = f(X_t, U_t)$. Let's assume at the certain moment of time the membership functions $\mu(\sigma_i)$ and $\mu(U_i)$ are given. In the Zadeh FST there find a sequence (U_0, U_1 … U_{n-1}), which maximizing the decision μ_D:

$$\mu_D(U_0, U_1, \ldots U_{N-1}) = \mu_0(U_0) \wedge \ldots \mu_{N-1}(U_{N-1}) \wedge \mu_N(U_N) .$$

The decision is: $U_n = \pi_n(x_n)$, where π_n is the accepted strategy, i.e., accepted rule of a choice of entrance signals e Un depending on realized Xn. After that the method of dynamic programming is applied to find Xn and maximizing (effective) decision.

Thus, Zadeh's fuzzy sets theory can became useful mathematical instrument in oil and gas production.

REFERENCES

[1] Zadeh L.A., Fuzzy Sets, *Information and Control*, 1965, v.8, p.338-353.

[2] Bellman R.E., Zadeh L.A. Decision-Making in a Fuzzy Environment//*Management Science*, 1970, vol.17, №4., p.141-164.

In: An Introduction to Fuzzy Sets
Editor: Valdemar F. Andersen
ISBN: 978-1-53618-012-1

Chapter 5

FUZZINESS-RANDOMNESS MODELING OF PLASMA DISRUPTION IN FIRST WALL OF FUSION REACTOR USING TYPE I FUZZY RANDOM SET

***Debabrata Datta*[1,*] *and Rashmi Bhardwaj*[2]**
[1]Senior Scientist-H, Head, Radiological Physics & Advisory Division, Bhabha Atomic Research Centre, CT&CRS Building, Anushaktinagar, Mumbai, India
[2]Professor of Mathematics, University School of Basic & Applied Sciences, Nonlinear Dynamics Research Lab, Guru Gobind Singh Indraprastha University, Delhi, India

ABSTRACT

TOKAMAK is a type of fusion reactor that use magnetic force to control plasma, fourth state of matter in which a separation between electrons and neutrons takes place by a pulsed high heat flux of magnitude

* Corresponding Author' E-mail: dbbrt_datta@yahoo.com.

2×10^5 MW/m^2. In TOKAMAK, heat flux gets injected into the plasma facing components due to plasma disruption. Melting and evaporation both takes place in the first wall of TOKAMAK type of fusion reactor due to which high heat load plasma facing components of the wall cracks. Ignorance of uncertainty (in this situation uncertainty is known as a mixture of probabilistic and epistemic uncertainty) is associated with the materials used for the design. Several researchers have performed experiments to study the melting and evaporation behaviour of the first wall of fusion reactor. In order to understand the initiation of melting and evaporation of plasma facing components, the first wall of a fusion blanket is modeled as 2-D rectangular slab with the surface facing plasma disruption by an applied heat flux. The rear surface of the slab is assumed as cooled by convection. The relevant parameters affecting the heat transfer during the early phases of heating as well as for large times are taken into account as uncertain due to their fuzziness and randomness. Fuzziness and randomness consistency principle is applied to model the uncertainty of the thermal response of the first wall. Fuzzy random set concept is very new to this kind of uncertainty modeling and therefore, this challenges the design of plasma facing components subject to simultaneous melting and evaporation. Finite Difference (Forward Time Central Space) based numerical solution with fuzzy random variables of the relevant parameters of the governing heat transfer equations is applied to obtain the temporal and spatial variation of temperature. The interpretation of results obtained are two folds: (a) Randomness uncertainty of the temperature of melting and evaporation of first wall of the plasma facing components is occurred at any alpha cut of the fuzziness, whereas (b) fuzziness uncertainty of the same components due to heat load is addressed at 95% confidence level by 2.5th and 97.5th percentile of the temperature. Water and Helium both are considered for cooling of the system. This chapter presents several challenges such as: (a) fuzzy random finite difference numerical method; (b) possibilistic uncertainty modeling; and (c) development of Fuzzy Wilks' theorem to model hybrid structure of randomness and fuzziness modeling.

Keywords: TOKAMAK, plasma disruption, randomness, fuzziness, uncertainty

1. INTRODUCTION

TOKAMAK is a Toroidal device which uses a powerful magnetic field to confine hot plasma in the shape of a torus. The *TOKAMAK* is one of several types of magnetic confinement devices being developed to produce controlled thermonuclear fusion power. In *TOKAMAK* reactor, by a nuclear fusion reaction of deuterium with tritium (D-T), plasma is produced and a large fraction (~80%) of the fusion energy is released. Approximately 14 MeV neutrons, charged particles and gamma radiation are also released during fusion reaction. It is required to slow down these neutrons by a structural material of large thickness. This structural material is known as blanket that surrounds the plasma. The kinetic energy of neutrons released is converted into heat energy having huge temperature. The first wall of the reactor is exposed to this heat flux which is corresponding to the energy released due to a plasma disruption. [1, 2]. The magnitude of the heat flux is predicted to be up to 2×10^5 MW/m^2. Since all of the emissions from reacting plasma may intersect the first wall, the same appears foremost as the region of concern. In summary, the first wall is melted, evaporated and cracked by this heat load. The damage observed in the first wall has a significant effect not only on the integrity but also on the life time of the plasma facing components (PFCs) of the first wall. In the worst case, failure of the reactor boundary would result the release of a significant mass of tritium from the reactor into the environment. In this context, a severe plasma disruption is one of the key issues for safety in fusion reactor.

Seki et al., [3, 4] have conducted the experimental studies on melting and evaporating characteristics of stainless steel, aluminum, zinc and graphite subjected to an extremely high heat flux up to 100 MW/m^2. They have developed an analytical code which is capable of treating melting and evaporation simultaneously. Many other authors [5, 6, 7] have presented an analytical model to analyze melting and evaporation behaviour during a plasma disruption. They have also shown that the results of their model were in reasonable agreement with those obtained from numerical analyses of the governing equations. Therefore, experimental studies anticipate that melting and evaporation of the first wall are due to heat load during plasma disruption. However the time at which plasma disruption will occur is not certain. Moreover, impurity present in the material of the first wall invites an uncertainty analysis of the thermo-physical phenomenon of melting and evaporation by an extremely high heat load due to a plasma disruption. We have two types of uncertainty. One is called as aleatory (random) uncertainty which is stochastic or probabilistic in nature. Aleatory uncertainty is irreducible. Other type of uncertainty is known as epistemic. Epistemic uncertainty is subjective in nature. Epistemic uncertainty is reducible and this is due to imprecise nature of the model parameters of the system of interest. In practice, imprecision is computed using fuzzy sets. A fuzzy set is defined as a paired number, such as $\tilde{A}=\{x,\mu\}$, where $x \in R^n$ and μ represents the membership value of crisp value x. Membership value can be represented as a convex bounded function such as triangular, trapezoidal, Gaussian, etc. Convex bounded membership function powers a fuzzy set as a fuzzy number which can be used to address the imprecision of model parameters of any physical or engineering system. In any engineering system, we always have an admixture of aleatory uncertainty and epistemic uncertainty. Aleatory uncertain parameters are characterized by randomness and epistemic uncertain parameters are characterized by fuzziness. Traditional Monte-Carlo statistical method is used to generate the realizations of aleatory uncertain parameters. Therefore, uncertainty of any engineering system is quantified by fuzziness-randomness modeling of the system.

This chapter will present the fuzziness-randomness modeling methodology of uncertainty analysis of melting and evaporation of the first wall of fusion reactor under plasma disruption. Uncertainty analysis due to the fuzziness and randomness of the material properties such as thermal conductivity, specific heat capacity and applied heat flux will be presented here. Since measurement system follows the normal distribution with a specified mean and standard deviation, random uncertainty (randomness) is estimated by normal distribution and fuzziness will be addressed by the triangular membership function. One can also address epistemic uncertainty by trapezoidal membership function. But question is which membership function is appropriate for which condition. If fuzziness is addressed by a phrase 'around mean', one should use triangular membership function and if fuzziness is addressed as an approximate between two numbers (lower and upper), fuzziness is addressed by trapezoidal membership function. Epistemic uncertainty of the melting and evaporation of the first wall will be evaluated in terms of the membership grades of the temperature profile of the first wall under the influence of melting and evaporation conditions.

2. The First Wall of Fusion Reactor

2.1. Physical Structure

The concept of first wall is accomplished by a relatively thick (~2 cm or so) structure that face plasma (Figure 1). Basically, this structure is a thermal mass with cooling tubes welded or brazed to the rear surface (far from the plasma face). There are three objectives for designing this type of thick wall. These objectives are: (a) protection of the cooling tubes from off-normal energy dumps due to plasma disruptions or thermal transients; (b) significant reduction of the thermal fluctuations seen by the cooling tubes for short pulse lengths, reduction of the alternating component of the thermal stress, and hence increasing of fatigue lifetime; and (c) anticipating presence of high fluxes of charge exchange neutrals to the first wall for substantial erosion of the first wall by likelihood of sputtering. Region of the fusion reactor is labeled as first wall blanket and the same is conceptualized geometrically as one dimensional rectangular slab for simplicity of modeling. The surface of the slab interfacing the plasma is subjected to an applied heat flux and the opposite (rear) surface is convectively cooled. Volumetric heating is due to neutron and gamma energy. Wide range of the time dependent behavior of the temperature profile of the first wall can be anticipated whereas steady-state analyses only provide the knowledge of the maximum temperature. In case of steady-state commercial power reactors (plasma burning for a longer time) the dominant mode of temperature behavior will be steady state but in case of experimental reactor, time profile of plasma burn being shorter (only a few seconds), a steady periodic temperature behavior will persist.

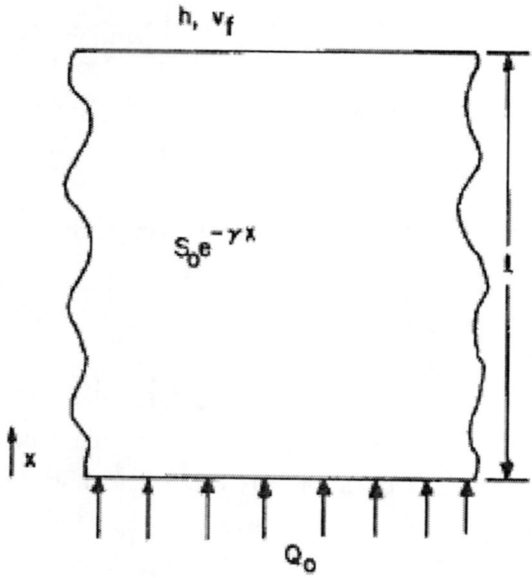

Figure 1. One dimensional slab geometry of the First Wall.

2.2. Mathematical Modeling of Heat Generation in First Wall

Physics of the system guides that the volumetric internal heat generation in the first wall can be assumed by a decaying exponential function. On the basis of this assumption, the governing equation representing one-dimensional time dependent heat conduction equation can be written as

$$\frac{\partial \theta}{\partial t} = \kappa \frac{\partial^2 \theta}{\partial x^2} + \frac{S_0}{\rho C} e^{-\gamma x} \varphi(t) \tag{1}$$

Initial and boundary conditions are as follows:

$$-\kappa\frac{\partial\theta}{\partial x}=Q_0\varphi(t),\ \ \mathrm{x}=0;\ \ -\kappa\frac{\partial\theta}{\partial x}=h(\theta-\theta_f),\quad \mathrm{x}=\mathrm{l}\ \ \text{and}\ \ \theta=\theta_0,\ \ t=0 \tag{2a}$$

$$Q_0\varphi(t),\ \ \mathrm{x}=0;\ \ -\kappa\frac{\partial\theta}{\partial x}=h(\theta-\theta_f),\quad \mathrm{x}=\mathrm{l}\ \ \text{and}\ \ \theta=\theta_0,\ \ t=0 \tag{2b}$$

and

$$\mathrm{x}=\mathrm{l}\ \ \text{and}\ \ \theta=\theta_0,\ \ t=0 \tag{2c}$$

where, θ is the temperature, S_0 and γ are parameters dependent on plasma conditions as well as blanket structure materials, Q_0 is the incident flux to the first wall, h is the heat transfer coefficient, κ represents the thermal diffusivity (cm²/s) and θ_f represents the mean fluid temperature. S_0 and γ are derived from results of neutronics analysis. Incident flux is assumed to be a certain percentage of the plasma output (~20%). Cooling is continuous during the absentee period of the plasma disruption.

We assume that the blanket structure is at some uniform temperature, θ_0 at time t = 0. Cooling behavior is characterized with the "on" and "off" state of the plasma disruption. The applied flux, $Q_0\varphi(t)$, represents a flux, Q_0, which is "on" for time, T_1 ($\varphi(t)=1$) and "off" for time, T-T_1 ($\varphi(t)=0$), with period T. The analytical solution to eq. (1) subject to eq. (2a-2c) is given by [8]

$$\theta=\theta_f+A+Bo+C \tag{3}$$

where,

$$A=\frac{Q_0 l}{k\ Bi}\left[1+Bi\left(1-\frac{x}{l}\right)\right], \tag{4}$$

$$Bo = \frac{S_0 l^2}{k\,B\,Bi}\left[1 + Bi\left(1 - \frac{x}{l}\right) + \left(\frac{Bi}{B} - 1\right)e^{-B} - \frac{Bi}{B}e^{-Bx/l}\right] \tag{5}$$

and

$$C = (\theta_0 - \theta_f)\sum_{n=1}^{\infty}\left\{\frac{2\sin\alpha_n}{\alpha_n} + \frac{2S_0 l^2}{k(\theta_0 - \theta_f)}\left[Bi\,e^{-B}\left(1 - \frac{B}{Bi}\right)\cos\alpha_n + B\right]\frac{1}{(\alpha_n^2 + B^2)\,\alpha_n^2} - D\right\}\times E \tag{6}$$

$$D = \frac{2Q_0 l}{k(\theta_0 - \theta_f)}\left(\frac{1}{\alpha_n^2}\right), \tag{7}$$

and

$$E = \frac{\alpha_n^2 + Bi}{\alpha_n^2 + Bi^2 + Bi}\left[\cos\left(\frac{\alpha_n x}{l}\right)\right]\exp\left(-\frac{\kappa\alpha_n^2 t}{l^2}\right) \tag{8}$$

The physical significance of the parameters Bi and B are as follows:

$Bi = Biot\ Modulus = \dfrac{hl}{k}$,

$B = \gamma\, l =$ dimensionless parameter and

$\tan\alpha_n = \dfrac{Bi}{\alpha_n}$

Setting the on period of plasma disruption as t = rT + t' (where 0 < t' < T_1 and r is large) and using Duhamel's theorem [9], we can write the steady state periodic variation in temperature that exists after the transient as

$$\theta = \theta_f + A + Bo - \Psi \tag{9}$$

where,

$$\Psi = \sum_{n=1}^{\infty}\left\{\frac{2S_0 l^2}{k\alpha_n^2(\alpha_n^2+B^2)}\left[Bi\, e^{-B}\left(1-\frac{B}{Bi}\right)\cos\alpha_n + B\right] + \frac{2Q_0 l}{k\alpha_n^2}\right\}\frac{\alpha_n^2+Bi^2}{\alpha_n^2+Bi^2+Bi}\times\Delta$$

$$\Delta = \left[\frac{\exp(\xi(T_1-t^{'})) - \exp(\xi(T-t^{'}))}{1-\exp(\xi T)}\right]\cos\left(\frac{\alpha_n x}{l}\right),$$

$$\xi = \frac{\kappa\alpha_n^2}{l^2}$$

The first two terms of eq. (9) are the steady state solution and the last term represents the steady, periodic variation of temperature with respect to the 'true' steady state.

In a similar way, by setting the 'off' period, $t = rT + T_1 + t^{''}$, we can write the steady periodic variation of temperature as

$$\theta = \theta_f + \sum_{n=1}^{\infty}\left\{\frac{2S_0 l^2}{k\alpha_n^2(\alpha_n^2+B^2)}\left[Bi\, e^{-B}\left(1-\frac{B}{Bi}\cos\alpha_n\right)+B\right] + \frac{2Q_0 l}{k\alpha_n^2}\right\}\times\Omega \tag{10}$$

where,

$$\Omega = \left[\frac{\exp(\xi(T-T_1-t^{''})) - \exp(\xi(T-t^{''}))}{1-\exp(\xi T)}\right]$$

Equations (9) and (10) provide the steady state periodic variation of temperature during 'on' and 'off' period of plasma disruption, which indicates that the thermal response of the first wall is a function of the time scale for plasma disruption and the plasma/wall interaction itself. As a result the wall is heated to its melting point and the melted material gets evaporated. Therefore, it is required to find the melting thickness and the time at which melting starts. As time passes, evaporation increases abruptly and the energy transported by thermal conduction decreases with increase of evaporation. Therefore, evaporation of the first wall is also presented here.

2.3. Melting of Solid Region (First Wall)

In case of melting of a solid temperature distribution exists both in the liquid and solid phase. Scenario for studying the melting behavior of a solid is like this: A constant heat flux 'q' is applied for a short period of time, T at one end of the slab which is initially at some uniform temperature below the melting point. The other end of the slab is at its initial temperature. Because of the sudden huge heat load the slab starts to melt. The problem is to determine the melting behavior of the slab through the uncertainty analysis. During the rise time to the melting temperature the heat transfer in the first wall is conduction or diffusion limited. Assuming there is no phase change, the slope of the first wall surface temperature matches the energy input and accordingly we have the boundary condition,

$$-k\frac{\partial \theta}{\partial x} = q = \frac{Q}{TA}, \quad x = 0 \tag{11}$$

Analytical solution of the eq. (1) subject to the boundary condition as given in eq. (11) is given by [8]

$$\theta - \theta_0 = \frac{2Q}{TA}\sqrt{\kappa t}\; ierfc\left[\frac{x}{2\sqrt{\kappa t}}\right] \tag{12}$$

where, A in eq. (12) signifies the area (cm^2) of the first wall subjected to disruption. The *ierfc* is related to the error function by

$$ierfc(x) = \int_{\infty}^{x} erfc(s)ds = \frac{1}{\sqrt{\pi}}\exp(-x^2) - x[1 - erf(x)], \quad erf(x) = \frac{2}{\sqrt{\pi}}\int_{0}^{x}\exp(-s^2)\,ds \tag{13}$$

2.4. Evaporation of Solid Region (First Wall)

As the temperature of the first wall increases due to the sudden dump of huge thermal load, the first wall starts to melt and both the liquid and solid phase of the metallic solid (first wall) appears. In other words the slab consists of distinct liquid and solid portions containing melted and unmelted material. Eventually vaporization of the first wall takes place. If we assume the thickness of the melting region as ξ(t) at time t, the boundary condition at this boundary of solid – liquid interface can be written as

$$-k_{liq}\,\frac{\partial\theta_{liq}}{\partial x}\Big|_{\xi} = -k_{solid}\,\frac{\partial\theta_{solid}}{\partial x}\Big|_{\xi} = -\rho\,L_m\,\frac{d\xi}{dt} \qquad (14)$$

$$\theta_{liq}(\xi,t) = \theta_{solid}(\xi,t) = \theta_m = \text{melting temperature and } \xi(0) = 0 \qquad (15)$$

The general solution to eq. (1) with the initial and boundary conditions as given in eq. (14) and (15) can be elsewhere found in [9]. The flux of the material vaporized can be obtained by using the Hertz-Knudsen-Langmuir equation [10] as

$$J_v = \alpha_v\,(P_s - P_0)\left(\frac{2\pi\,M\,K_B\,\Delta\theta(0,t)}{N_0}\right)^{-1/2} \qquad (16)$$

where, J_v = net vaporization flux (atoms/area-time),

α_v = evaporation coefficient = 1.0

$\Delta\theta(0,t)$ = surface temperature of the first wall

P_s = saturation pressure of the vapor at the surface temperature

P_0 = partial pressure of the vapor

M = molecular weight of the vapor species

N_0 = Avogadro's number

K_B = Boltzmann's constant

For the convenience of uncertainty analysis here we have assumed $P_0 = 0$ which corresponds to evaporation into a vacuum. The saturation pressure corresponding to the surface temperature $\Delta\theta(0,t)$ can be computed by

$$P_s = \exp[(A/\Delta\theta(0,t)) + E] \tag{17}$$

where, A and E are constants depend on the type of the material of the first wall. The thickness of the material that is vaporized at every time step can be expressed by

$$\xi_v = \frac{J_v M}{\rho N_0}\Delta t = \frac{\alpha_v \exp[(A/\Delta\theta(0,t)) + E]}{\rho N_0}\left(\frac{2\pi\, M\, K_B \Delta\theta(0,t)}{N_0}\right)^{-1/2} \Delta t \tag{18}$$

2.5. Alpha Cut of a Fuzzy Variable and Fuzzy Random Variable

Fuzzy variable is generally used to model epistemic uncertainty which is based on ambiguity, imprecision and lack of information. In order to implement, we express the membership function of fuzzy variable in terms of alpha cut [9, 10]. Alpha cut of a fuzzy variable is defined as an interval defined over $\mathbf{R}^n$. Horizontal representation of a fuzzy variable is expressed by its alpha cut. Mathematical representation of alpha cut of a fuzzy variable can be written as $\mu[\alpha] = \{x \in X | \mu(x) \geq \alpha\}$ [11]. On the contrary strict alpha cut of the fuzzy variable is represented as = $\{x \in X | \mu(x) > \alpha\}$. Since alpha cut of a fuzzy variable is an interval, interval arithmetic is used to estimate the epistemic uncertainty. However, simulation of alpha cut value by Monte Carlo approach converts fuzzy variable into a fuzzy random variable. But strictly speaking, concept of fuzzy random variable is the better approach for expressing fuzziness-randomness of an uncertain variable. Fuzzy random variable is defined as that variable where parameters of randomness (e.g., mean and standard deviation for Gaussian distribution) are expressed in terms of a fuzzy variable with specified membership function (either, triangular or trapezoidal). The set of such fuzzy random variable is known as fuzzy random set. Here, we have addressed only type- I fuzzy random set because fuzziness part of the variable is considered as type I fuzzy set. In the simulation of fuzzy random variable, we assume that the mean and standard deviation, then in fuzzy random variable, the mean and standard deviation are considered as fuzzy variable with either triangular or trapezoidal membership function.

2.5.1. Fuzzy Random Variable

Mathematical definition of fuzzy random variable with corresponding fuzzy probability space can be represented in the following way:

Let $G_0(R)$ denote the set of all bounded closed fuzzy numbers, i.e., if $\overline{A} \in G_0(R)$, then $\tilde{r}$ satisfies

$$\{x \mid x \in R,\ \overline{A}(x) = 1\} \neq \phi;$$

For every $\alpha \in (0,1]$,

$\tilde{A}_\alpha = \{x \mid x \in R, \tilde{A}(x) \geq \alpha\} = [A_\alpha{}^-, A_\alpha{}^+]$ is a finite closed interval on R,

where $R = (-\infty, \infty)$

2.5.1.1. Definition

Fuzzy random variable on the probability space (Ω, A, P) is defined as a mapping $\tilde{a} : \Omega \rightarrow G_0(R)$, if for every $b \in B$, following equation is satisfied

$$\{w \mid w \in \Omega, \tilde{a}_\alpha(w) \cap B \neq \phi\} \in A, \text{ for every } \alpha \in [0, 1],$$

where B represents the Borel subsets on R and

$\widetilde{a}_{\alpha}(w) = \{x \mid x \in R, \widetilde{a}(w)(x) \geq \alpha\}$ is the level set of $\widetilde{a}(w)$.

2.5.1.2. Theorem

If $a_{\alpha}^{-}(w)$ and $a_{\alpha}^{+}(w)$ are two random variables such that $a_{\alpha}^{-}(w) \leq a_{\alpha}^{+}(w)$ everywhere on Ω then there exists a closed interval random number $\widetilde{a}(w) = [a_{\alpha}^{-}(w), a_{\alpha}^{+}(w)]$ on the probability space (Ω, A, P).

2.5.1.3. Definition

If $\widetilde{a}, \widetilde{b} \in FR(\Omega)$, we can define

(1) $\widetilde{a} \leq \widetilde{b}$ if and only if $\widetilde{a}_{\alpha}(w) \leq \widetilde{b}_{\alpha}(w)$ for every $w \in \Omega$ and for any $\alpha \in [0,1]$.

(2) $\widetilde{a} < \widetilde{b}$ if and only if $\widetilde{a}_{\alpha}(w) < \widetilde{b}_{\alpha}(w)$ for every $w \in \Omega$ and for any $\alpha \in [0,1]$.

2.5.1.4. Definition

If o be an algebraic operation on $F_0(R)$ (o may be "+", "-", "*", "/", etc), then the same algebraic operation o on FR(Ω) can be defined by $(\widetilde{a} \; o \; \widetilde{b})(w) \underline{\underline{\Delta}} \; \widetilde{a}(w) \; o \; \widetilde{b}(w)$ for any $w \in \Omega$, where $\widetilde{a}, \widetilde{b} \in$ FR(Ω).

For standard arithmetic operation of closed interval numbers, one can see [6-11]. Therefore, it is obvious that $R \subset G_0(R) \subset FR(\Omega)$, and $R \subset R(\Omega) \subset FR(\Omega)$, where $R(\Omega)$ is the set of all random variables on (Ω, *A*, *P*).

2.6. Fuzzy Random Variable Valued Gaussian Function

We have introduced the structure of fuzzy random variable-valued Gaussian function into this section. Fundamental properties of fuzzy random variable valued Gaussian function are presented. It is mandatory to discuss about the following definitions and theorems.

2.6.1. Definition

The mapping $f:\widetilde{D}\rightarrow FR(\Omega)$ $(\widetilde{x}\mapsto f(\widetilde{x}))$ is called a fuzzy random variable-valued function defined on $\widetilde{D}$, the domain of definition for function $f(\widetilde{x})$ and represents any nonempty subset of FR(Ω) and $\widetilde{x}$ is called the independent variable. It is also understood that $f(\widetilde{x})$ is also a fuzzy random variable in FR(Ω) for every $\widetilde{x}\in\widetilde{D}$.

The definition and fundamental properties of fuzzy random variable-valued exponential function is described in detail elsewhere in [12]. Similar strategy is followed to present the fuzzy random variable valued Gaussian function.

If $\widetilde{a}\in FR(\Omega)$ be a given fuzzy random variable then

$$f_1(\widetilde{x})=\widetilde{a}.\widetilde{x},\ \text{for any } \widetilde{x}\in\widetilde{D}$$

$$f_2(\widetilde{x})=\widetilde{a}+\widetilde{x},\ \text{for any } \widetilde{x}\in\widetilde{D}$$

are two fuzzy random variable-valued functions defined on $\widetilde{D}$.

It is understood that we may structure maps from $\widetilde{D}$ to *FR(Ω)* by all sorts of methods. We shall use the extension principle to define the fuzzy random variable valued Gaussian function.

2.7. Fuzzy Random Variable Valued Gaussian Function

$$\exp(-\widetilde{x}^2/2)$$

2.7.1. Lemma

Suppose that f(x) $(x \in R)$ is an ordinary real valued continuous function, $\widetilde{a} \in F_0(R)$. Defining, $f(\widetilde{a}) \triangleq \bigcup_{\alpha \in (0,1]} \alpha\, f(\widetilde{a}_\alpha)$, we can write

1. for any $\alpha \in (0,1]$,
$(f(\widetilde{a}))_\alpha = f(\widetilde{a}_\alpha) = [\wedge f(x),\ \vee f(x)], \qquad x \in \widetilde{a}_\alpha$
2. $f(\widetilde{a}) \in F_0(R)$

For proof of the lemma 2.7.1, one can follow [13]. It can be pointed here that $f(\widetilde{a})$ is also a bounded closed fuzzy number in $F_0(R)$.Therefore, using this definition for an exponential function, we can write the following expression

$$\exp x = e^x, (x \in R = (-\infty, \infty)),\ \widetilde{x} \in FR(\Omega),$$
$$\exp \widetilde{x} \triangleq \bigcup_{\alpha \in (0,1]} \alpha \exp(\widetilde{x}_\alpha),$$

where, $\exp(\widetilde{x}_\alpha) = \{\exp x \mid x \in \widetilde{x}_\alpha = [x_\alpha^-, x_\alpha^+]\}$, $\alpha \in (0,1]$ and $\exp \widetilde{x}$ is called as the fuzzy random variable valued exponential function with respect to base e. One of the important property of $\exp \widetilde{x}$ at its α-level representation is $(\exp \widetilde{x})_\alpha = \exp \widetilde{x}_\alpha = [\exp x_\alpha^-, \exp x_\alpha^+]$, for any $\alpha \in (0,1]$ [13]. For every $x \in FR(\Omega)$ using lemma 2.7.1 and this property we can say that x_α^- and x_α^+ are two random variables in the fuzzy probability space for any $\alpha \in (0,1]$.

2.7.2. Definition

Let, $\exp(-\widetilde{x}^2/2) = e^{-\widetilde{x}^2/2} = \psi(\widetilde{x})$, (where $x \in R = (-\infty, \infty)$) be an ordinary Gaussian function. For a fuzzy random variable, $\widetilde{x} \in FR(\Omega)$, we can define

$$\exp(-\widetilde{x}.\widetilde{x}/2) \underline{\underline{\Delta}}\, \alpha \exp(-\frac{\widetilde{x}_\alpha.\widetilde{x}_\alpha}{2}),$$

where

$$\exp(-\frac{\widetilde{x}_\alpha.\widetilde{x}_\alpha}{2}) = \left\{\exp(-\frac{x^2}{2}) \mid x \in \widetilde{x}_\alpha = [x_\alpha^-, x_\alpha^+]\right\}, \alpha \in [0,1].\ \exp(-\frac{\widetilde{x}.\widetilde{x}}{2})$$

is called the fuzzy random variable valued Gaussian function.

Let, $\widetilde{y} = \frac{\widetilde{x}.\widetilde{x}}{2}$, Therefore, $\exp(-\frac{\widetilde{x}.\widetilde{x}}{2}) = \exp(-\widetilde{y})$. So, we can write

$$\widetilde{y}_\alpha = \frac{1}{2}(\widetilde{x}.\widetilde{x})_\alpha = \frac{1}{2}[\widetilde{x}_\alpha].[\widetilde{x}_\alpha] = \frac{1}{2}[x_\alpha^-, x_\alpha^+].[x_\alpha^-, x_\alpha^+] = \frac{1}{2}[x_\alpha^-.x_\alpha^-, x_\alpha^+.x_\alpha^+]$$

2.7.3. Theorem

(1) $(\exp(-\widetilde{y}))(\omega) \in F_0(R)$ for every $\omega \in \Omega$

(2) $(\exp(-\widetilde{y}))_\alpha = \exp(-\widetilde{y}_\alpha) = [\exp(-y_\alpha^-),\ \exp(-y_\alpha^+)]$, for any $\alpha \in (0,1]$

(3) $\exp(-\widetilde{y}) \in FR(\Omega)$

For proof, readers can follow the reference [13].

2.7.4. Theorem

Let, $\widetilde{x}_1$, $\widetilde{x}_2 \in FR(\Omega)$. Then,

$$\exp(-[(\widetilde{x}_1.\widetilde{x}_1)/2 + (\widetilde{x}_2.\widetilde{x}_2)/2]) = \exp(-\frac{\widetilde{x}_1.\widetilde{x}_1}{2}).\exp(-\frac{\widetilde{x}_2.\widetilde{x}_2}{2})$$

2.7.5. Proof

Since for any $\alpha \in (0,1]$,

$$(\tilde{x}_1^2+\tilde{x}_2^2)_\alpha=(\tilde{x}_1^2)_\alpha+(\tilde{x}_2^2)_\alpha=\{[\tilde{x}_1^-,\tilde{x}_1^+]_\alpha.[\tilde{x}_1^-,\tilde{x}_1^+]_\alpha+[\tilde{x}_2^-,\tilde{x}_2^+]_\alpha.[\tilde{x}_2^-,\tilde{x}_2^+]_\alpha\}$$
$$=\{[\tilde{x}_1^-.\tilde{x}_1^-+\tilde{x}_2^-.\tilde{x}_2^-]_\alpha,[\tilde{x}_1^+.\tilde{x}_1^++\tilde{x}_2^+.\tilde{x}_2^+]_\alpha\}$$

Hence,

$$(\tilde{x}_1^2+\tilde{x}_2^2)_\alpha^-=\tilde{x}_{1\ \alpha}^-\tilde{x}_{1\ \alpha}^-+\tilde{x}_{2\ \alpha}^-\tilde{x}_{2\ \alpha}^-$$
$$(\tilde{x}_1^2+\tilde{x}_2^2)_\alpha^+=\tilde{x}_{1\ \alpha}^+\tilde{x}_{1\ \alpha}^++\tilde{x}_{2\ \alpha}^+\tilde{x}_{2\ \alpha}^+$$

Thus by theorem 2.7.1, we can write the followings:

$$(\exp(-\frac{\tilde{x}_1^2+\tilde{x}_2^2}{2}))_\alpha=\exp(-(\frac{\tilde{x}_1^2+\tilde{x}_2^2}{2})_\alpha)=[\exp(-(\frac{\tilde{x}_1^2+\tilde{x}_2^2}{2})_\alpha)^-,\exp(-(\frac{\tilde{x}_1^2+\tilde{x}_2^2}{2})_\alpha)^+]$$
$$=[\exp(-(\frac{\tilde{x}_1^2}{2})_\alpha)^-.\exp(-(\frac{\tilde{x}_2^2}{2})_\alpha)^-,\ \exp(-(\frac{\tilde{x}_1^2}{2})_\alpha)^+.\exp(-(\frac{\tilde{x}_2^2}{2})_\alpha)^+]$$
$$=[\exp(-(\frac{\tilde{x}_1^2}{2})_\alpha)^-,\exp(-(\frac{\tilde{x}_1^2}{2})_\alpha)^+][\exp(-(\frac{\tilde{x}_2^2}{2})_\alpha)^-,\exp(-(\frac{\tilde{x}_2^2}{2})_\alpha)^+]$$
$$=[\exp(-\frac{\tilde{x}_1^2}{2})_\alpha][\exp(-\frac{\tilde{x}_2^2}{2})_\alpha]=[\exp(-\frac{\tilde{x}_1^2}{2})\exp(-\frac{\tilde{x}_2^2}{2})]_\alpha,\quad \text{for any}$$

$\alpha\in(0,1]$

Consequently, $\exp\left(-\frac{\tilde{x}_1^2+\tilde{x}_2^2}{2}\right)=\exp\left(-\frac{\tilde{x}_1^2}{2}\right)\exp\left(-\frac{\tilde{x}_2^2}{2}\right)$

3. Uncertainty Modeling

Uncertainty analysis of the temperature of the first wall due to plasma disruption is carried out using eq. (12). As the material of the first wall melts due to plasma disruption, it is always required to obtain the uncertainty quantification of the melting thickness and eq. (15) is utilized for this purpose. It is also known that as the melting progress due to the increase of the temperature of the first wall, a certain amount of the melted portion also evaporates at every time step. In view of this physical fact, uncertainty analysis of the thickness of the material vaporized is also carried out using eq. (18). Random uncertainty of the thermal conductivity and the fuzziness of density of the material of the first wall are used as input parameters for uncertainty quantification of the temperature, melting thickness and the vaporized thickness of the first wall. Alpha cut concept of fuzzy variable is mixed with random variable. In our work fuzziness of the density is transformed into randomness by fuzzy alpha cut technique. Alpha cut of a fuzzy variable is an interval and assuming that interval as the bounds of a uniform distribution, traditional Monte Carlo is executed to generate the random space. The same operation is repeated for all alpha cut values and then by probabilistic inversion method, probability distribution of density parameter is constructed.

4. Results and Discussions

The material of the first wall of the fusion reactor is assumed as SS-316. The input (deterministic) parameters of this material used in the analysis are tabulated in Table 1. The uncertainty of both the input parameters is assumed as uniformly distributed and presented in Table 2. The traditional Monte Carlo method has been used to assess the uncertainty of the surface temperature of the first wall due to huge thermal load during plasma disruption. The same method also has been adopted to evaluate the uncertainty of the melting and vaporization of the first wall. Latin hypercube sampling technique has been used to generate the realization of the uncertain parameters.

Time of observations used in the uncertainty analysis of the surface temperature of the first wall is varying in the range from 0 - 50 millisecond. Frequency plot of the surface temperature corresponding to time of observation after 10 msec is as shown in Figure 2.

Table 1. Deterministic input parameters for material SS-316

Specific heat (J/kg-oK)	700
Total energy deposited to the first wall during disruption (J)	25000
Disruption time (msec)	10
Area of the first wall facing plasma (m2)	1
Latent heat of vaporization (J/kg)	6270 E+3
Constant, A	-46627
Constant, E	22.13
Melting Temperature (oK)	1672

Table 2. Probabilistic input parameters for material SS-316

Parameters	Lower limit	Mean value	Upper limit
Thermal conductivity (W/m-oK)	25	30	35
Density (kg/m3)	6000	7000	8000

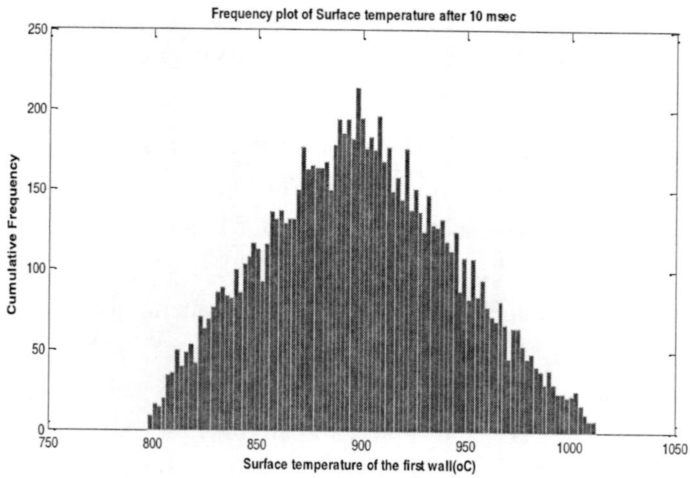

Figure 2. Histogram of surface temperature of the first wall after 10 millisecond.

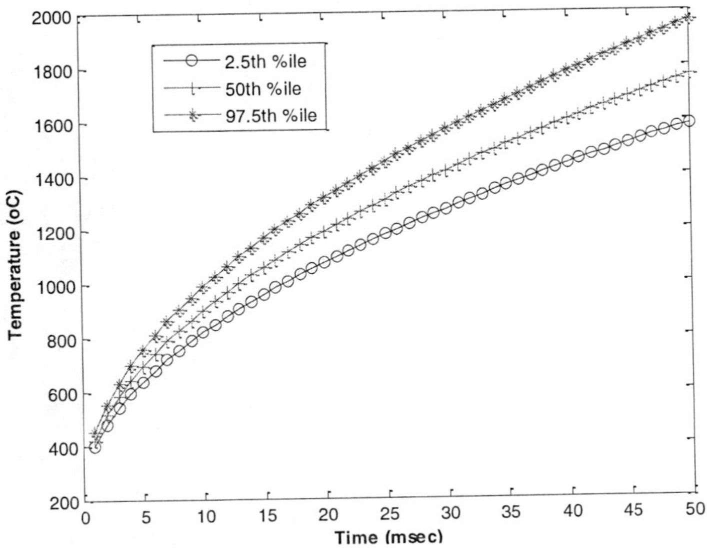

Figure 3. Uncertainty plot of the surface temperature of the first wall at various time steps.

Figure 3 represents the results of uncertainty of the surface temperature of the first wall (x = 0) due to deposition of the total energy of 25000 J on unit area of the first wall for a time of 10 millisecond during plasma disruption. Uncertainty of the output (surface temperature) at every time step is expressed by 2.5^{th}, 50th and 97.5^{th} percentiles of the surface temperature of the first wall. Uncertainty of the melting thickness of the first wall due to plasma disruption is estimated using eq. (15). The melting temperature of the SS-316 is known as 1399°C and accordingly the uncertainty width (2.5^{th} and 97.5^{th} percentiles) of the surface temperature of the first wall at time 45 millisecond is estimated which is the denominator of the right hand side (RHS) of eq. (15). The numerator $\Delta\theta(x_m, t = 45 \text{ msec}) = \theta_m$ (melting temperature) of the right hand side of eq. (13) is taken as 1399°C. Computation of the RHS of eq. (15) results 0.4268 (corresponding to 97.5^{th} percentile of $\Delta\theta(0, t = 45 \text{ msec})$ and 0.5270 (corresponding to 2.5^{th} percentile of $\Delta\theta(x_m, t = 45 \text{ msec})$. The lookup table of the 'ierfc(ζ)' provides the corresponding value of ζ as 0.15 and 0.038 respectively. By using eq. (15) with known value of all the parameters (Q, T, A and $\Delta\theta(0, t = 45)$ and utilizing these ζ values the corresponding melting thickness is estimated as

0.1mm and 20 mm. Assuming the physical thickness of the first wall as 2 cm one can easily say that 50% of the material will melt and that will initiate after a time of 45 millisecond of plasma disruption. In a similar way by substituting the respective values of the constants and the uncertainty of the surface temperature of the first wall in eq. (18) uncertainty of the thickness of the material vaporized at each time step is estimated as [2.5e-7 mm, 8.57e-5 mm]. Upper bound of the variation of the thickness vaporized at each time step is as shown in Figure 4.

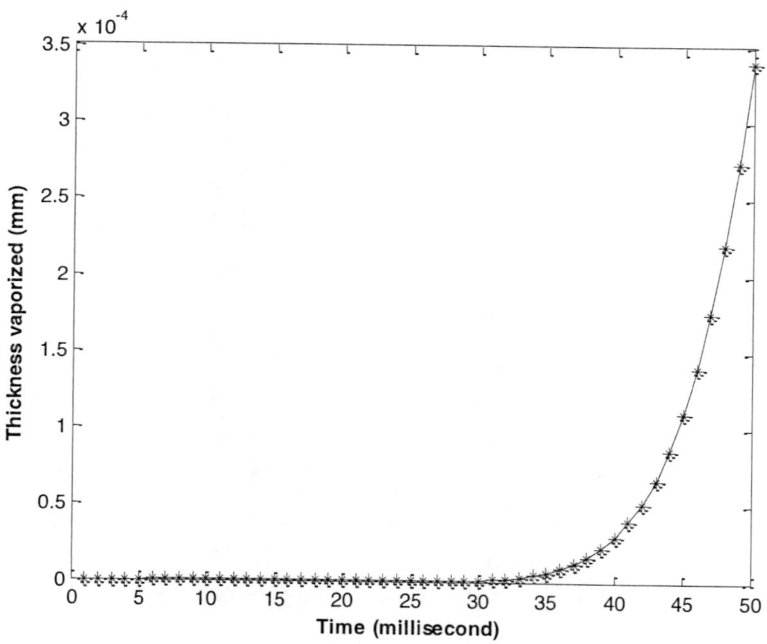

Figure 4. Variation of the upper bound of the thickness of the material vaporized at each time step.

Conclusion

Uncertainty analysis of melting and evaporation of the first wall of fusion reactor under plasma disruption is presented. Temperature profile of the first wall based on its physics study is presented analytically in eq. (3).

Steady periodic variation of temperature of the first wall for 'on' and 'off' state of plasma disruption are described in eq.(9) and eq.(10). Traditional Monte Carlo simulation method is applied to carry out the uncertainty analysis. Randomness of the material properties such as thermal conductivity and the density is considered as input uncertainty parameters. Basically the aleatory uncertainty of the surface temperature of the first wall facing plasma, melting and evaporation thickness of the first wall are quantified. Uncertainty band of all these assessments is expressed using 2.5th and 97.5th percentiles of the respective random output (e.g., surface temperature, melting and evaporation thickness).

REFERENCES

[1] *ITER Concept Definition.*

[2] Akiba, M., et al.: J. *At. Energy Soc.* Jpn, 35(5), 447, 1991.

[3] Seki, M., et al.: *J. Fusion Energy*, 5(3), 181, 1986.

[4] Seki, M., et al.: *JAERI-M* 01-090, 1990.

[5] Sanzo, D.: Nucl. Eng. Des./Fusion, 4, 191, 1987.

[6] Cohen, S.A., et al.: *Nucl. Fusion*, 21, 233, 1981.

[7] Hassanein, A.M., et al.: *Nucl. Eng. Des.*/Fusion, 1307, 1984.

[8] Carslaw, H.S. and Jaeger, J.C., *Conduction of Heat in Solids* (Oxford University Press, London, 1959).

[9] Timothy Ross, *Fuzzy Logic with Engineering Applications*, Third edition, Wiley Publications, 2010.

[10] Dubois.D & Prade.H., Operations of fuzzy numbers, *Int. J. Systems Sci.* 9(6),613-626,1978.

[11] Dubois.D & Prade.H., Fuzzy sets and systems, *Theory and Applications*, 1980.

[12] Zhang Yue and Wang Guangyuan, *The Theory of Fuzzy Stochastic Dynamic Systems* (Science Press, Beijing, 1993).

[13] Qiao Zhong, Zhang Yue and Wang Guangyuan, Fuzzy random variable-valued exponential function, logarithmic function and power function, *Fuzzy Sets and Systems,* 99 (1998), pp. 311-324.

In: An Introduction to Fuzzy Sets
Editor: Valdemar F. Andersen

ISBN: 978-1-53618-012-1

Chapter 6

APPLICATION OF A STANDARD FUZZY ARITHMETIC METHOD

Sangeeta Kumari*, PhD
Department of Civil Engineering,
NIT Jamshedpur, Jharkhand, India

ABSTRACT

A standard fuzzy arithmetic method is used for solving fuzzy equations and optimization of fuzzy objectives. The fuzzy variable of the equation is fuzzified using a fuzzy set. A fuzzy set is defined as a membership function with a degree of membership varying from zero to one where zero represents non-membership, one represents full membership and values between zero and one represent partial membership. The vertical axis of the fuzzy set is decomposed into several intervals. Each interval is solved separately using the interval theory. The resultant intervals are recomposed according to degree of membership and resultant fuzzy set is obtained. The resultant fuzzy sets are defuzified to get the crisp values. The fuzzy arithmetic method is applied to a case study of the reservoir operation in which fuzzy storage continuity and fuzzy soil moisture balance equations are solved and fuzzy model objective of the

* Corresponding Author's E-mail: Sangeeta.iitm@gmail.com.

minimizing crops moisture deficits is obtained. The method is found quite effective and easy to apply in solving complex equations.

Keywords: Fuzzy arithmetic, Reservoir operation, Storage balance, Soil moisture balance, Crop moisture deficits

INTRODUCTION

The fuzzy set theory has been used for handling uncertainties in the variables which may occur due to randomness, lack of adequate data, imprecisions and subjectivity. The fuzzy set is defined as a membership function with degrees of membership varying from zero to one where zero indicates none membership to one indicates full membership and values between zero to one indicate partial memberships. The membership of a fuzzy set can be classified as triangular, trapezoidal, Gaussian, etc. The fuzzy arithmetic has been used to solve complex fuzzy equations in which fuzzy numbers are decomposed into several intervals and interval arithmetic are applied to solve the equations and then output intervals are recomposed to get resultant fuzzy numbers. In this chapter, the fuzzy arithmetic method is applied to obtain optimal release policies from an irrigation reservoir by minimising the crop soil moisture deficits and by maintaining the reservoir storage continuity and soil moisture balance.

METHODS

Reservoir Operation Model

The fuzzy state dynamic programming model (Kumari and Mujumdar 2015) for an irrigation reservoir is formulated in which reservoir storage and soil moistures of the crops are fuzzy variables. The backward recursive equation for the model is given as:

$$\bar{f}_t^n(k,i,M) = \underset{\{feasible\ l\}}{Min}\left[\bar{G}(k,i,l,M,t) \oplus \textstyle\sum_j P_{ij}^t \otimes \bar{f}_{t+1}^{n-1}(l,j,N)\right] \quad (1)$$

where $\bar{f}_t^n(k,i,M)$ is the minimum system performance measure up to the end of the last period T when n stages are remaining for the feasible combinations of storage class interval k, inflow class interval i, and soil moisture class intervals M. P_{ij}^t is transitional probability of inflow during the period $t+1$ in the class interval j given that the inflow during the period t is in the class interval i. $\bar{G}(k,i,l,M,t)$ is the fuzzy objective of the model for combination of initial storage class interval k, inflow class interval i, final storage class interval l, soil moisture class intervals M and period t. The fuzzy objective of the model at period t is a summation of daily fuzzy objectives of the model within the period.

The objective function of the model is minimization of the crops moisture deficits which is given as:

$$\bar{G}(k,i,l,M,d) = Min \textstyle\sum_{c=1}^{C} ky_c^d \otimes \left[1 \ominus \overline{ET}_{a,d}^c \oslash ET_{max,d}^c\right] \quad (2)$$

where d is a day, c is a crop, ky_c^d is the yield factor of the crop c at day d, $\overline{ET}_{a,d}^c$ and $ET_{max,d}^c$ are actual and potential evapotranspiration respectively, and C is the total number of crops.

The storage continuity equation for the reservoir is given by

$$\bar{S}_l^{t+1} = \bar{S}_k^t \oplus Q_i^t \ominus \bar{R}_{kil}^t \ominus \bar{E}_{kl}^t \; S_{min} \leq \bar{S}_k^t \leq S_{max} \quad (3)$$

where $\oplus$ and $\ominus$ are fuzzy addition and subtraction respectively, $\bar{S}_k^t$ is the initial reservoir storage belongs to the initial storage class k. $\bar{S}_l^{t+1}$ is the final reservoir storage belongs to the final storage class l, Q_i^t is the reservoir inflow belongs to the inflow class i, $\bar{R}_{kil}^t$ is the reservoir release belongs to the initial storage class k, the inflow class i, and the final storage class l. S_{min} and S_{max} are the dead storage capacity and the maximum storage capacity of the reservoir respectively. $\bar{E}_{kl}^t$ is the evaporation loss in the reservoir.

The soil moisture balance equation is given as:

$$\bar{\theta}_{Fd} = \frac{[(\bar{\theta}_m^d \otimes D_d)\oplus \overline{IRA}_d \oplus RAIN_d \ominus \overline{ET}_a^d]}{D_d} \quad Z_w \leq \bar{\theta}_m^d \leq Z_f \tag{4}$$

where ⊗ and ⊘ are fuzzy product and division respectively, D_d is the crop root depth, $\bar{\theta}_{Fd}$ and $\bar{\theta}_m^d$ are the final and the initial soil moisture respectively, Z_w and Z_f are permanent wilting point and the field capacity respectively, $\overline{IRA}_d$ is the depth of irrigation applied, and $RAIN_d$ is the daily rainfall.

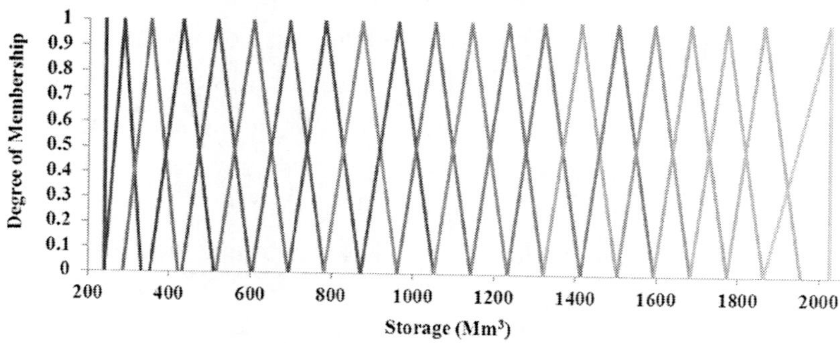

Figure 1. Fuzzification of Reservoir Storage (Mm^3).

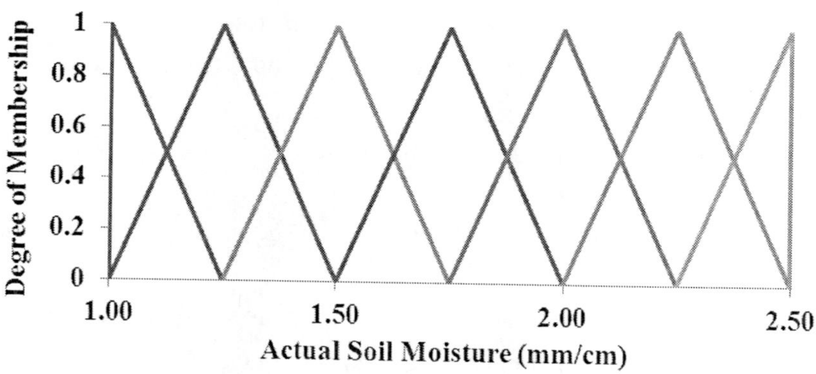

Figure 2. Fuzzification of Soil Moisture (mm/cm).

The storage and soil moistures of the crops are fuzzified using a triangular membership considering overlapping between adjacent intervals as shown in Figures 1 and 2 respectively.

A STANDARD FUZZY ARITHMETIC METHOD

This method is used to perform the arithmetic operations such as addition, subtraction, multiplications between two or more fuzzy numbers. It consists of the following steps:

Step 1. Decomposition of Fuzzy Numbers

In the first step, the fuzzy interval [0, 1] of the μ-axis (vertical axis) is divided into m intervals of length $\Delta\mu = {}^{1}/_{m}$ and the discrete values μ_j of the $(m+1)$ levels of membership are then given by

$$\mu_j = j/m \, , j = 0,1, \ldots \ldots \ldots, m \qquad (5)$$

Then the fuzzy number $\tilde{p}_i$ are decomposed into alpha-cuts, leading to the decomposed representations as:

$$\mathcal{P}_i = \left\{x_i^{(0)}, x_i^{(1)}, \ldots \ldots \ldots, x_i^{(m)}\right\}, i = 1,2, \ldots \ldots, n, \qquad (6)$$

where $\mathcal{P}_i$ consists of the $(m+1)$ intervals.

$$x_i^{(j)} = \left[a_i^{(j)}, b_i^{(j)}\right] = cut_{\mu_j}(\tilde{p}_i), a_i^{(j)} \leq b_i^{(j)}, j = 1,2, \ldots \ldots, m \qquad (7)$$

Step 2. Application of an Interval Arithmetic (Moore 1966)

The fuzzy rational expression $\tilde{q} = f(\tilde{p}_1, \tilde{p}_2, \ldots \ldots, \tilde{p}_n)$ is computed by evaluating the interval-valued counterparts separately at each level of membership μ_j.

$$z^{(j)} = f\left(x_1^{(j)}, x_2^{(j)}, \ldots \ldots, x_n^{(j)}\right), j = 0, 1, \ldots \ldots \ldots, m \qquad (8)$$

The evaluation of these interval rational expressions is performed by successive execution of elementary interval arithmetic according to the definitions of the basic operations as follows.

$$\left[a_1^{(j)}, b_1^{(j)}\right] \star \left[a_2^{(j)}, b_2^{(j)}\right] = \left[\underbrace{\min(\mathcal{H}^{(j)})}_{a^{(j)}}, \underbrace{\max(\mathcal{H}^{(j)})}_{b^{(j)}}\right], * \in \{+, -, \times, \div\} \quad (9)$$

$$\mathcal{H}^{(j)} = \left\{a_1^{(j)} * a_2^{(j)}, a_1^{(j)} * b_2^{(j)}, b_1^{(j)} * a_2^{(j)}, b_1^{(j)} * b_2^{(j)}\right\} \quad (10)$$

Step 3. Recompositions of the Output Intervals

As a result of the application of interval arithmetic, the value of the fuzzy rational expression is available in its decomposed representation.

$$O = \left\{Z^{(0)}, Z^{(1)}, \dots\dots\dots, Z^{(m)}\right\} \quad (11)$$

By recomposing the intervals $Z^{(j)}, j = 0,1, \dots\dots, m$ of O according to their levels of membership μ_j, the value $\bar{q}$, the output fuzzy number can be obtained.

FUZZY RANKING METHOD

It is used to compare two or more fuzzy numbers and is based on the extension principle for ordering fuzzy numbers, say $\bar{p}_1, \bar{p}_2, \dots\dots\dots\dots \bar{p}_n$. $\mathcal{A}(\bar{p}_i)$ is the degree to which $\bar{p}_i$ is ranked as the greatest fuzzy number which can be defined for each $i \in \mathcal{N}_n$ as:

$$\mathcal{A}(\bar{p}_i) = sup \min_{k \in \mathcal{N}_n} p_k(r_k) \quad (12)$$

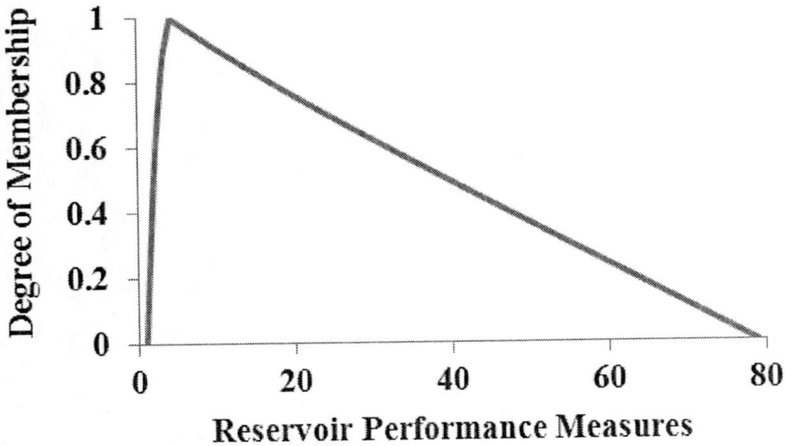

Figure 3. Resultant Fuzzy Performance Measure of the Reservoir.

where the supremum is taken over all vectors $\langle r_1, r_1, \ldots . r_n \rangle \in \mathcal{R}^n$ such that $r_i \geq r_j$ for all $j \in \mathcal{N}_n$. $\mathcal{N}_n = \{1, 2, \ldots \ldots \ldots, n\}$ is the set of the natural numbers, $\mathcal{R}^n$ is the set of all fuzzy numbers of length n.

The standard fuzzy arithematic method is used to solve reservoir operation model and fuzzy ranking method is used to compare the fuzzy numbers. The resultant fuzzy performance measures obtained is shown in Figure 3.

Thus, It can be concluded that a standard fuzzy arithmetic can be used to solve complex equations of the reservoir operation models. Further, it is quite useful in handling real time situations.

REFERENCES

Kumari, S. and Mujumdar, P. (2015). "Reservoir Operation with Fuzzy State Variables for Irrigation of Multiple Crops". *ASCE Journal of Irrigation and Drainage Engineering,* Volume: 141, Issue: 11, 121121121Article Number: 04015015, pp. 1 - 13.

Moore, R. E. (1966). *Interval analysis*. Prentice-Hall, Englewood Cliffs, NJ, USA.

BIOGRAPHICAL SKETCH

Sangeeta Kumari

Affiliation: Department of Civil Engineering, NIT Jamshedpur, Jharkhand-831014, India.

Education: PhD in 2017 in Water Resources Engineering from Indian Institute of Science Bangalore, India.

Research and Professional Experience: Involved in water resources modelling for 12 years and Currently working as an assistant professor since last 1.5 years.

Professional Appointments:

1. Assistant Manager (July 2006 to March 2011) in Fluidyn Software and Consultancy (P) Limited, Bangalore, India.
2. Assistant Professor (June 2018 to Present) at NIT Jamshedpur, Jharkhand, India.

Honors: NIL

Publications from the Last 3 Years:

[1] Kumari Sangeeta (2019). "Fuzzy-set-based real-time reservoir operation using genetic algorithm". *ISH Journal of Hydraulic Engineering,* 1 - 8, Taylor & Francis, DOI: 10.1080/09715010.2019.1604181.

[2] Kumari, Sangeeta and Mujumdar, P. P. (2017). "Fuzzy-Set based System Performance Evaluation of an Irrigation Reservoir System". *ASCE Journal of Irrigation and Drainage Engineering*, Volume: 143, Issue: 5, Article Number: 04017002, pp. 1 - 14, 10.1061/ (ASCE) IR.1943-4774.0001155 (Impact Factor - 1.364, Citations: 04).

INDEX

L

M

N

O

P

Q